CMS WORKSHOP LECTURES

VOLUME 5

COMPUTER APPLICATIONS TO X-RAY POWDER DIFFRACTION ANALYSIS OF CLAY MINERALS

THE CLAY MINERALS SOCIETY

J. R. WALKER, SERIES EDITOR

R. E. FERRELL, Jr., EDITOR-IN-CHIEF

CMS WORKSHOP LECTURES
VOLUME 5

COMPUTER APPLICATIONS TO X-RAY POWDER DIFFRACTION ANALYSIS OF CLAY MINERALS

by

D. L. Bish, Los Alamos National Laboratory, Los Alamos, NM
A. Blum, U.S. Geological Survey, Menlo Park, CA
D. D. Eberl, U. S. Geological Survey, Boulder, CO
R. Jones, University of Hawaii, Honolulu, HI
H. Malik, University of Hawaii, Honolulu, HI
D. R. Pevear, Exxon Production Research, Houston, TX
R. C. Reynolds, Jr., Dartmouth College, Hanover, NH
J. F. Schuette, Exxon Production Research, Houston, TX
J. R. Walker, Vassar College, Poughkeepsie, NY

Editors

R. C. Reynolds, Jr
Department of Earth Sciences
Dartmouth College
Hanover, NH 03755

J. R. Walker
Department of Geology
Vassar College
Poughkeepsie, NY 12601

Published by

The Clay Minerals Society
P.O. Box 4416
Boulder, CO 80306

Copyright © 1993 The Clay Minerals Society

All rights reserved. No part of this book may be reproduced by any mechanical, photographic, or electronic process or in the form of a photographic recording, nor may it be stored in a retrieval system, transmitted, or otherwise copied for public or private use without written permission from the publisher.

For information on this and other volumes in the CMS Workshop Lecture Series, write The Clay Minerals Society, P.O. Box 4416, Boulder, CO, 80306, U. S. A.

Citations of articles or chapters in this volume are properly referenced as follows:

Author (1993) Title: in CMS Workshop Lectures, Vol. 5, Computer Applications to X-ray Powder Diffraction Analysis of Clay Minerals, R. C. Reynolds, Jr. and J. R. Walker, eds., The Clay Minerals Society, Boulder, CO, 000-000

Library of Congress Catalog Card Number
ISBN - 1-881208-06-0

TABLE OF CONTENTS

Preface	vi
The Clay Minerals Society	viii
An Introduction to Computer Modeling of X-ray Diffraction Patterns of Clay Minerals: A Guided Tour of NEWMOD© J. R Walker	1
Inverting the NEWMOD© X-ray Diffraction Forward Model for Clay Minerals Using Genetic Algorithms D. R. Pevear and J. F. Schuette	19
Three-dimensional X-ray Powder Diffraction from Disordered Illite: Simulation and Interpretation of the Diffraction Patterns R. C. Reynolds, Jr.	43
Studies of Clays and Clay Minerals Using X-ray Powder Diffraction and the Rietveld Method D. L. Bish	79
Illite Crystallite Thickness by X-ray Diffraction D. D. Eberl and A. Blum	123
A Computer Technique for Rapid Decomposition of X-ray Diffraction Instrumental Aberrations from Mineral Line Profiles R. C. Jones and H. U. Malik	155

PREFACE

The advent of high-speed desk-top computers has made it possible to solve many crystallographic problems which heretofore required long hours of calculating. More importantly, many problems can be addressed that were simply inaccessible by any means thirty years ago. Some of the programs described in this volume complete more calculations in minutes than an individual could accomplish over a lifetime even if every second of that lifetime was spent with pencil and paper in hand. And the necessary computing tools are widely available at prices comparable to two or three of the best slide rules in the 1950's. We do routinely what could not have been imagined by scientists of a few generations ago. And this is just the beginning--we will do Rietveld analyses some day on laptop computers that will be disposed of when their batteries run down.

Beginning with the MOD series in the late 60's, the modeling of basal reflections became almost *de riguer* in X-ray powder diffraction studies of clay minerals. As with most things, as computers become faster and more powerful, fewer simplifying assumptions needed to be made so mathematical models could be made more complete and rigorous. Widespread use of programs based on the Rietveld method has allowed the analysis of three-dimensional structures of clays as long as no disorder is present, and this method or variations of it may some day make single crystal methods obsolete.

The present volume, produced to accompany a short course by the same name at the 1993 Annual Meeting of the Clay Minerals Society in San Diego, CA, presents the state-of-the-art in computing as applied to several types of powder diffraction studies of clay minerals. The first three chapters address computer applications which model the characteristics of an X-ray diffraction pattern from user-specified atomic/crystal structure parameters. The second three chapters present applications which extract characteristics of a clay structure from experimental data.

An introduction to many of the concepts and terms used in this volume is given by Walker in the context of a discussion of the workings of NEWMOD©, a program familiar to many workers in the clay field. The chapter also gives hints about how to use NEWMOD© effectively and efficiently. More importantly, Walker encourages researchers to get to know the NEWMOD© code, and to be fearless in changing things to suit their laboratory conditions and special situations.

An example of the lengths to which the code can be modified is given by Pevear and Schuette who use a genetic algorithm to generate and modify parameters for modeling powder diffraction patterns from mixtures of clay minerals common in shales. The genetic algorithm does an excellent job modeling even very complicated synthetic diffraction patterns created using NEWMOD©, giving hope that its models of natural samples are also correct. The genetic algorithm is certainly faster than a human researcher, and the author's approach may be the definitive method for quantitative analysis of clay mineral mixtures in the future.

Modeling of three-dimensional diffraction effects from illites is the subject of the chapter by Reynolds, who shows convincingly that many of the possible structural variations on the illite structure probably occur in nature. These structures require a complicated model because they include both *cis*-vacant and *trans*-vacant forms, interstratifications of these, and interlayer rotations of 60, 120, 180, 240, and 300°. The program, however, can be easily executed by any of the faster desk-top computers. It is up to the investigator to make informed choices of starting parameters, but Reynolds provides a detailed discussion of what to look for, and where, to find clues about the probable structural details that describe a sample.

The Rietveld method, which allows one to refine crystal structure information (such as lattice parameters and atomic coordinates) from powder diffraction data, appears to be an ideal method to study fine-grained materials such as clay minerals which are not available as large single crystals. Bish discusses the advantages of the Rietveld method when applied to clay minerals, but is careful to discuss at length the many cases in which the method cannot be used. The most common cause of failure of a Rietveld refinement is the presence of disorder such as semi-random or turbostratic stacking, however, the nature of the Rietveld method is such that an answer is always produced. Bish discusses examples in which the answer looks much better than it really is when the result is scrutinized in detail. He also enumerates the many possible applications of Rietveld analysis, including quantitative analysis, lattice parameter refinement, peak shape analysis, and crystal structure refinement of complex mixtures.

Clay peaks are, by nature, broader than those of minerals which posses long range three-dimensional order, and the source of this broadening is the subject of the final two papers in the volume. The broadening is partly a function of the nature of the sample and partly a function of the instrument on which the analysis was made. Sample-related effects are related to discontinuities in the mineral structure which break the crystallite down into smaller coherent diffracting domains. A nice feature of X-ray powder diffraction peaks is that their shapes contain information on the nature of these discontinuities. NEWMOD© allows the investigator to model the differences in peak shape due to particle size and to the presence of defects. The Rietveld method allows one to separate strain broadening from particle-size broadening. The Warren-Averbach method is used by Eberl and Blum to measure crystallite thickness of illites. They compare their measurements against thicknesses measured by various microscopic and chemical techniques, and point out that, because X-ray powder diffraction measures many more grains than a microscopist could measure in a lifetime, a diffraction technique which makes the measurements quickly and easily would result in a great savings of time and an increase in the amount of data which could be generated.

Modeling the contribution of the instrument to diffraction peak broadening is usually done by comparing an experimental profile with a profile from a crystalline material with "infinitely" large diffracting domains. However, Jones and Malik point out that any material contains defects which make it not an "infinite" crystal, so they attempt to model the instrumental signature using profile-fitting. They make the assumption that the pure mineral profile can be described by a symmetrical pseudo-Voight function, and then model the remaining asymmetry (mostly on the low-angle side) with one or more peaks representing the contribution of the instrument. Comparison of their results with peak shapes from a diffractometer on a synchrotron radiation source, suggests that their method holds great promise.

The papers included in this volume represent the state-of-the-art in a science which is rapidly changing. It is hoped that the work presented here will inspire others to continue to develop and use new computer methods for analyzing X-ray powder diffraction patterns of clays and other fine-grained materials. There are still plenty of unanswered questions, and the computing power is just around the corner.

J. R. Walker

R. C. Reynolds, Jr.

THE CLAY MINERALS SOCIETY

The Clay Minerals Society was organized in 1963 to stimulate research and to disseminate information relating to all aspects of the science and technology of clays and other fine-grained minerals. It sponsors the annual Clay Minerals Conferences where research and invited papers are presented in technical sessions and special symposia and organizes field trips to important occurrences of clays in all parts of North America and to industrial sites of clay production and application. In conjunction with its annual meetings, workshops are held on technical subjects of interest to clay researchers and technologists.

The Clay Minerals Society publishes bimonthly *Clays and Clay Minerals*, which is one of the leading international journals in the field of clay science. *Clays and Clay Minerals* presents the latest scientific investigations in all areas of the field and from all parts of the world, along with timely review articles and announcements of new publications on clays and other fine-grained minerals. Its half-tone illustrations of electron micrographs and other photographs are among the best scientific reproductions to be found.

The Society also sponsors a Source Clays project designed to provide homogeneous clay samples for research and teaching purposes and offers clay-related publications to its members at a discount over publisher's list prices,

Its multidisciplinary membership includes agronomists, surface chemists, physicists, geologists, mineralogists, geochemists, materials scientists, soil scientists, crystallographers, sedimentologists, economic geologists, colloid chemists, ceramists, rheologists, petroleum engineers, and geotechnical engineers, thereby offering members the opportunity to exchange ideas and results with fellow researchers having widely different backgrounds and expertise.

For information on joining the Society and subscribing to *Clays and Clay Minerals*, write to:

The Clay Minerals Society
Society Office
P.O. Box 4416
Boulder, CO 80306
USA

Phone: (303) 444 - 6405
FAX: (303) 444 - 2260

AN INTRODUCTION TO COMPUTER MODELING OF X-RAY POWDER DIFFRACTION PATTERNS OF CLAY MINERALS: A GUIDED TOUR OF NEWMOD©

J.R. WALKER

CONTENTS

Introduction	2
Basic Intensity Calculation	3
The Layer Structure Factor	3
The Lorentz-Polarization Factor	6
The Interference Function	6
The Frequency Factor	8
The Complete Intensity Equation	9
Additional Considerations	10
Peak Shape Modeling	10
Particle-Size Versus Defect Broadening	10
Mixed-Layer Broadening	12
Modeling Phases Not Included In NEWMOD©	14
Concluding Remarks	15
Acknowledgments	16
References Cited	16

AN INTRODUCTION TO COMPUTER MODELING OF X-RAY POWDER DIFFRACTION PATTERNS OF CLAY MINERALS: A GUIDED TOUR OF NEWMOD©

J.R. WALKER

Department of Geology and Geography
Vassar College
Poughkeepsie, NY 12601

INTRODUCTION

Discussion of the calculation of powder X-ray diffraction (XRD) profiles offers an excellent opportunity to introduce many of the concepts of XRD modeling which will be basic to understanding the succeeding chapters in this volume. This introductory chapter will discuss in detail the theory underlying the one-dimensional diffraction algorithm used by the NEWMOD© software series (Reynolds, 1985). Attention will also be given to some of the more specialized aspects of the NEWMOD© series especially the statistics used to describe interstratifications. The NEWMOD© series is familiar to many researchers in the clay field, and calculating XRD profiles with it is common in many investigations. However, the details of the calculations may not be familiar to all. In this paper, NEWMOD© variable names are given in all capitals, for example SIGSTAR or LOWN, and are included to help interested persons navigate through the source code.

The intensity of X-rays diffracted by infinitely large crystals, neglecting constants, is given by

$$I(hkl) = |F(hkl)|^2 * LP \qquad (1)$$

where F is the structure factor which describes diffraction from an arrangement of atoms, and LP is the combined Lorentz and polarization factor which describes the contribution of several important experimental conditions to diffraction. The absolute value of the structure factor is used in Eq. (1) because the phase of diffracted X-rays is not known, so the magnitude of F can be determined, but not its sign. Both F and LP can be described mathematically and, therefore, are amenable to incorporation into algorithms to calculate model X-ray diffraction patterns. The structure factor is defined in terms of the spacing of the *hkl* planes contributing to diffracted intensity at a given angle, the so-called "Bragg" angle, as well as the identity of the atoms which make up those planes. Because diffraction at angles not corresponding to a given interplanar spacing, or *d*-value, is minimal in a large crystal, the intensity equation is solved only for the Bragg angles. Results of such a calculation are the familiar "stick figures" showing intensity at the Bragg positions such as can be produced from information in the JCPDS files.

In the real world, however, intensity is distributed over some finite angular range even for perfect crystals (if such exist) due to instrumental broadening. Some computer programs, such as POWD10 (Smith et al., 1983), approximate instrumental broadening by distributing intensity over an angular range according to a statistical function such as a Lorentzian or Gaussian, and a user-specified peak width. For clay minerals, however, peaks are also broad because the diffracting planes are short, the crystallites are thin, and mixed-layering is common (Reynolds, 1989b). The

Guided Tour of NEWMOD©

formulation of the structure factor, therefore, is modified to describe diffraction as a continuous function of 2θ, to give the layer structure factor (G), and the effects of finite crystallite size and mixed layering are included in the interference function (Φ). These modifications allow one to describe scattering which occurs between the Bragg positions. Eq. (1) is therefore modified to

$$I(\theta) = |G(\theta)|^2 * LP * \Phi . \qquad (2)$$

One of the distinguishing features of NEWMOD© is its ability to model mixed-layered clay systems through the use of Markovian statistics. A frequency descriptor (σ) is used to quantify the frequency of a given layer sequence within the interstratified system, and Eq. (1) is further modified to

$$I(\theta) = |G(\theta)|^2 * LP * \Phi * \sigma. \qquad (3)$$

All of quantities on the right side of Eq. (3) except LP are summed over the different layer types in the system. This is the general form of the intensity equation used in NEWMOD©.

The component parts of Eq. (3) will be discussed in the following sections. The discussion draws heavily from the work of Reynolds (1965; 1967; 1968; 1976; 1980; 1983; 1985; 1986; Bethke and Reynolds, 1986). The reader is referred to Moore and Reynolds (1989) for a lucid summary of these concepts, and to Reynolds (1989a; 1989b) for detailed discussions and references. The discussion presented herein is limited to one-dimensional diffraction from crystallites which are very similar in the X-Y plane and which differ primarily along the Z axis. The general principles, however, can be applied to three-dimensional diffraction in the study of polytypes, order-disorder relationships and other details of the three-dimensional structures of clay minerals (Reynolds, this volume).

BASIC INTENSITY CALCULATION

The Layer Structure Factor

The one-dimensional structure factor for diffracted intensity from a set of *00l* planes in a centrosymmetric structure at a specified Bragg angle can be expressed as

$$F(00l) = \sum_j n_j f_j \cos(2\pi l\, z_n/c) \qquad (4)$$

where n in the number of j-type atoms, f is the scattering power of j-type atoms which has been corrected for thermal vibrations, l is the order of the reflection, z_n is the displacement of the layer of j-type atoms from the center of symmetry, and c is the unit cell height. Non-centrosymmetric structures such as kaolinite and serpentine require the addition of a sine series summation, a complication which will not be treated here (see Reynolds, 1985; Moore and Reynolds, 1989).

The scattering power of an atom varies with diffraction angle and with oxidation state, and must be calculated for each atom type at each angle. NEWMOD© uses a polynomial to calculate scattering power (Wright, 1973). Included in NEWMOD© are Si^{2+}, $Al^{1.5+}$, Fe^{2+}, Mg^{2+}, K^+,

Ca^{2+}, Sr^{2+}, Na^+, C, H, O^{1-}, and O. Partial valences for Si, Al and O reflect the fact that these atoms form bonds which are in-part ionic and in-part covalent.

Because the calculation is one-dimensional, atom positions are given only in terms of their distance from the center of the structure in a direction normal to the layer. This amounts to the same thing as projecting all of the atoms onto that normal as shown in Figure 1. NEWMOD© uses a small number of sets of z_n values because of the essential similarity of the layers of most clays: dioctahedral and trioctahedral 1:1 and 2:1 silicate layers, Mg-OH and Al-OH octahedral layers, a hydrated interlayer cation, a fixed interlayer cation, and ethylene glycol. One and two layers of water or glycol can also be modeled. It is important to note that adjustments of layer thickness in NEWMOD© (D001A and D001B) only change the size of the interlayer space and do not affect the interatomic distances shown in Figure 1.

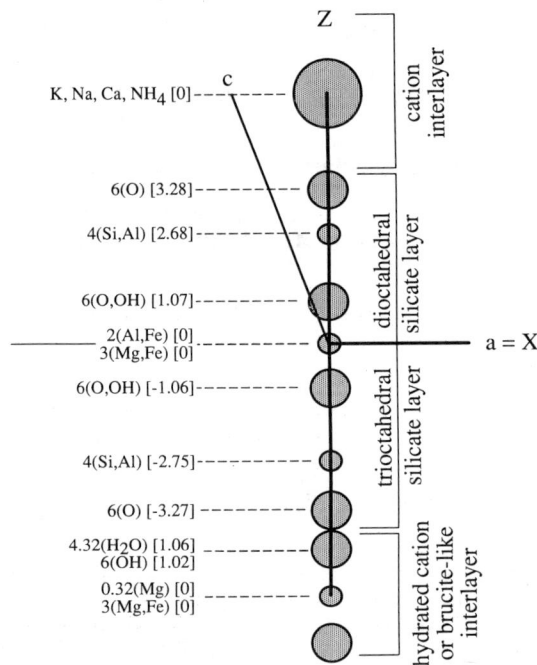

Figure 1. One-dimensional atom locations, projected on to Z, used to describe various layer types in NEWMOD©. From top to bottom the layer types are interlayer cation, dioctahedral silicate layer, trioctahedral silicate layer, and hydrated cation or brucite-like interlayer.

Guided Tour of NEWMOD©

In order to change the structure factor (F), which is only valid for discrete values of l, to the layer structure factor (G), which is continuous with respect to diffraction angle, Eq. (4) must be converted to units of θ. This conversion is made using the Bragg Law $n\lambda = 2d \sin\theta$. Since $l = n$, the Bragg Law may be rearranged to give $l = 2d \sin\theta/\lambda$. Substituting this into Eq. (4), and clearing the fractions, yields

$$G(\theta) = \sum_j n_j f_j \cos(4\pi z_n \sin\theta/\lambda). \tag{5}$$

A plot of $G(\theta)$ for a trioctahedral silicate layer is shown in Figure 2.

Equation (5) is the essence of the layer structure factor as used by NEWMOD© for centrosymmetric structures, however, to describe diffraction from layers in an interstratified system, it must be separated into two parts, one describing the silicate layers (G_S) and one describing the interlayers (G_A or G_B for a two component system). In this way, scattering from different centrosymmetric layers can be included, and end effects are avoided. The use of the two parts of G is discussed in greater detail in the section below that treats the complete intensity expression.

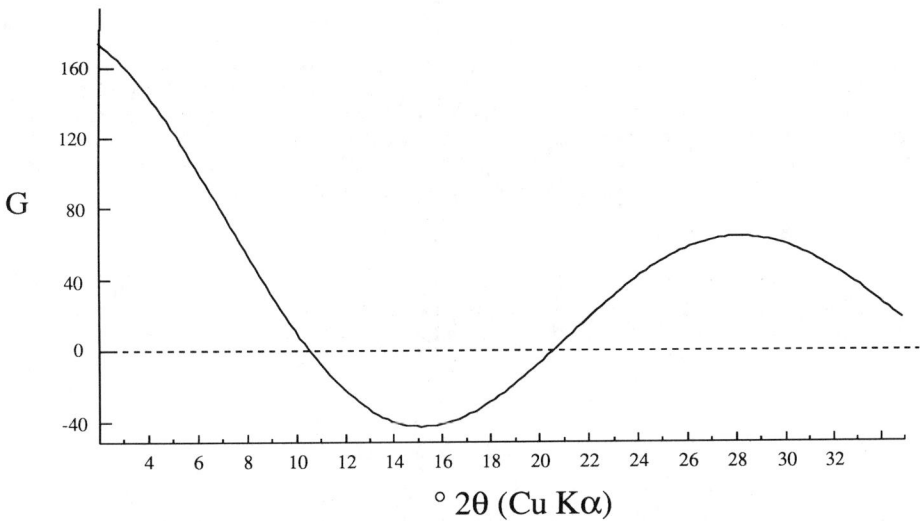

Figure 2. Layer structure factor, a continuous function of diffraction angle, for a trioctahedral silicate layer.

The Lorentz-Polarization Factor

The Lorentz-polarization factor (*LP*) combines several quantities related to the geometry of diffraction. The polarization factor, $(1+\cos^2 2\theta)/2$, describes changes in intensity due to polarization of the X-ray beam during scattering. The Lorentz factor is more complex than the polarization factor, involving the irradiated volume of the sample and the orientation of crystallites in the sample. The irradiated volume is described by $1/\sin\theta$ which is the so-called "single crystal" Lorentz factor. In a powder sample, however, the orientation of the crystallites also is important to the final magnitude of scattering. For a powder sample, the orientation part of the Lorentz factor is described by the powder ring distribution factor (Ψ) (Reynolds, 1986) which accounts for scattering from those crystallites oriented to produce diffraction at a given angle. If all grains in a sample are perfectly oriented parallel to the sample surface, Ψ approaches a constant and the single crystal Lorentz factor is the only part which varies with diffraction angle; for randomly oriented grains, Ψ is equal to $1/\sin\theta$ and the complete Lorentz factor is $1/\sin^2\theta$. Oriented mounts of clay minerals fall somewhere in between. so calculation of the powder ring distribution factor is extremely important for correct modeling of experimental diffraction patterns. For the powder ring distribution factor to effectively model experimental patterns, the orientation of crystallites in the sample must be known (σ^* or SIGSTAR, measured by a rocking curve), and the axial divergence of the Soller slit configuration (SOLLER1 and SOLLER2) must be correct for the instrument on which measurements were performed. The complete *LP* used by NEWMOD© is

$$\frac{1 + \cos^2 2\theta}{\sin\theta} \Psi \qquad (6)$$

which is solved for each angular increment in the experimental range. It is important that the instrumental conditions used by NEWMOD© be as close as possible to those of the original experiment so that the relative intensity of diffraction peaks will reflect the structure and composition of the mineral in question. The single crystal ($\sigma^* = 0$) and random powder ($\sigma^* = 45$) *LP*'s are plotted in Figure 3; solving Eq. (6) for a natural sample will produce a curve lying between the two curves depending on the orientation of grains in the sample.

The Interference Function

For small or thin crystals, or crystals containing more that one layer type, significant scattering occurs between the Bragg positions. This scattering can be modeled in terms of the interference function which describes the scattering of radiation by the diffraction grating formed by the clay layers

$$\Phi = \frac{\sin^2(2\pi ND \sin\theta/\lambda)}{N\sin^2(2\pi D \sin\theta/\lambda)} \qquad (7)$$

where N is the number of lines in the grating and D is their spacing. For clays, N is equivalent to the number of unit cells stacked in a coherently diffracting array, and D is the interplanar spacing, or d(001). In this form, the interference function gives maxima, all of which have the same integrated intensity, at the appropriate Bragg positions and $N - 2$ ripples between the maxima (Figure 4). Using a range of equally weighted values for N will smooth the ripples into a continuous curve (Figure 4). The breadth of peaks produced by Eq. 7 depends on the magnitude

Guided Tour of NEWMOD©

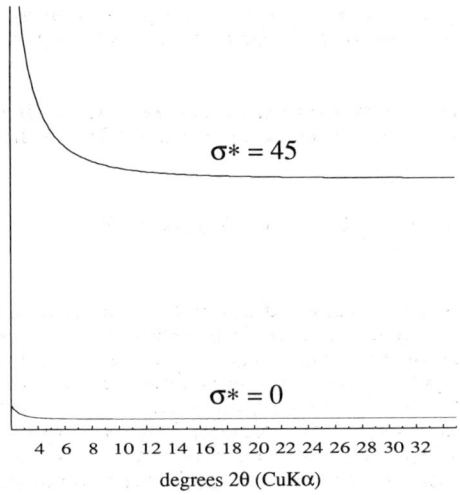

Figure 3. Limiting curves of the Lorentz-polarization factor as a function of preferred orientation of mineral grains in a sample mount.

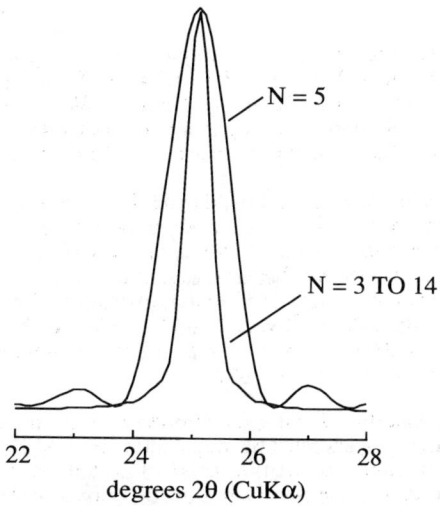

Figure 4. Interference function for chlorite 004 for crystallites which are all the same thickness ($N = 5$), and for crystallites ranging from 3 to 14 layers thick.

of N: small values of N give broad peaks and vice versa. Depending on the slopes of G and LP in the vicinity of a peak, broadening due to small values of N can cause peaks to migrate upslope from their Bragg positions.

The interference function can also be expressed as a Fourier series (James, 1965), in order to calculate the diffraction effects of any two layers within a coherently diffracting array,

$$\Phi = \frac{N}{2} + \sum_{n=1}^{n=N-1}(N-n)\cos(4\pi R_s(\sin\theta)/\lambda)) \tag{8}$$

where N is the number of layers per crystal, n is the number of interlayer spacings separating two layers under consideration, and R_s is the distance between the two layers. There are many advantages to expressing the interference function as a Fourier series as in Eq. (8). Because the summation is taken over all layer pairs within a crystallite, the interaction between any two individual layers can be considered. To model a mixed-layered clay, this ability has obvious advantages, because the individual layers may have very different frequencies of occurrence and layer structure factors, both of which can be included within the summation as discussed below in the section on the complete diffraction equation. In both pure and mixed-layered systems, the ability to calculate layer interactions is also important for modeling stacking disorder. To model the effects of a random distribution of stacking defects resulting in "defect broadening" of diffraction peaks, the weighting factor $(N-n)$ in Eq. (8) may be modified by a probability term describing the occurrence of defects as discussed below.

The Frequency Factor

The frequency of layer sequences within the crystal must be quantified to describe the scattering from layers in an interstratified mineral. The statistics of interstratification are specified in NEWMOD© using the Reichweite nomenclature (Jadgozinski, 1949). Specification of a Reichweite initiates a series of calculations resulting in a frequency factor (σ) which describes the probability of occurrence of a given layer sequence, or array, within a crystallite.

The frequency factor is made up of several parts: the number of times a given array will fit into a crystallite of thickness N, the proportion of each layer type in the array, and the junction probabilities (Reynolds, 1980) describing the possible sequences of layers within the array. The number of frequency factors increases with N because as particles become larger, more and larger arrays can fit in them. For the example of random interstratification, if $N = 4$, there are 16 $(= 2^4)$ arrays with 4 layers, 8 $(= 2^3)$ with 3, 4 $(= 2^2)$ with 2, and 2 $(= 2^1)$ with 1. However, if $N = 10$, there are 2^{10} arrays with 10 layers, 2^9 with 9, etc. A separate frequency factor must be calculated for each array (Moore and Reynolds, 1989, p. 316-7).

The number of junction probabilities within each frequency factor increases with ordering in a mixed-layered system because more layers must be included for a complete description. For $R = 0$ and $R = 1$, nearest neighbor junctions are all that need to be considered, however, for $R > 1$ ordering, next nearest neighbors, and neighbors removed by two or more layers, need to be included. Calculation of the junction probabilities is handled recursively (Bethke and Reynolds, 1986), a strategy which speeds up the algorithm considerably.

The frequency factor (σ) modifies the interference function given by Eq. (8) by including the frequency with which layers occur within an array

$$\Phi_{jk} = P + \sum_{n=1}^{n=N-1} \sigma_{jk}(N-n)\ \cos(4\pi R_s \sin\theta/\lambda) \qquad (9)$$

where the subscripts j and k represent the layer or interlayer types which terminate the array and P describes the occurrence of different layer or interlayer pairs. For silicate layer pairs $P = N/2$, for interlayer pairs $P = (P_i(N-1))/2$ where P_i is the probability of occurrence of an i-type interlayer, and for mixed layer pairs, such as a silicate layer and a interlayer or two different interlayers, $P = 0$. Note that the summation is over all layer pairs in the array, each with its own frequency factor σ_{jk}, and that a different Φ_{jk} is calculated for each array.

The Complete Intensity Equation

By combining the different factors discussed above, the complete intensity equation for a two component interstratification consisting of two types of interlayers A and B and a silicate skeleton S, and assuming all layers are centrosymmetric, may be expressed as

$$I(\theta) = LP * EX * (G_A G_B \Phi_{AB} + G_A G_S \Phi_{AS} + G_B G_S \Phi_{BS} + G_{AA}^2 \Phi_{AA} + G_{BB}^2 \Phi_{BB} + G_{SS}^2 \Phi_{SS}) \qquad (10)$$

where the Φ_{jk} terms are calculated according to Eq. (9), and the subscripts AA, BB, SS, AB, AS, BS refer to arrays terminating with the layer types represented by the letters. Asymmetric structures require a more complex expression (Reynolds. 1985; Moore and Reynolds, 1989).

The term *EX* in Eq. (10) includes all corrections to total intensity due to sample and instrumental characteristics, excluding the contribution of preferred orientation to *LP*. Some of these corrections vary with diffraction angle whereas others are constants. Most important of the quantities which are angle dependent is the sample length (SAMPLEN) correction. Sample length, coupled with the angular divergence of the anti-scatter slits (DIVSLIT or THETACOMP for a theta-compensating slit) causes a loss in intensity at low angles when the footprint of the X-ray beam irradiates an area larger than the sample causing some of the incident radiation to miss the sample. Quantities which do not vary with diffraction angle, including the quartz reference intensity (QTZREF) and the mass absorption coefficient (MUSTAR), are used along with an empirical constant (EMPCON) to adjust absolute intensities. Correction for some experimental variables are not included in NEWMOD©, including sample thickness, for which one needs to know the mass and area of material in the sample mount, instrumental broadening due to imperfect resolution of the Kα1α2 doublet, and inaccuracies in *d*-values due to sample displacements. If these variables are critical to the analysis at hand, they must be corrected for in the experimental pattern prior to modeling.

Of the many variables describing the physical characteristics of each sample and each analysis, some are specific to the laboratory involved whereas others are characteristic of the sample under scrutiny. For those variables in the former category, which can include GONRAD (the goniometer radius), SOLLER1, SOLLER2, DIVSLIT, THETACOMP, SAMPLEN, QTZREF, and MUSTAR, the values may be changed in the file DEFVAL to reflect normal operating procedures in the investigator's laboratory. Quantities in the latter category, which might include SAMPLEN, MUSTAR, SIGSTAR, D001A, D001B, HIGHN, LOWN, PROP OF Ns ALL ONE, are best changed from within the main program so that their new values will be recorded in the output file and included on any plots which are made.

Equation (10) is solved for each increment over the specified angular range, the resultant being a calculated diffraction profile of a mixed-layered clay with the composition and ordering parameters specified, analyzed according to the experimental conditions specified. If the parameters input by the investigator are correct for the mineral under study, then the match between the experimental and calculated patterns should be very good. It is important to note, however, that there is not always a unique solution, and good fits can often be obtained using a variety of input parameters. NEWMOD© does not evaluate the statistical "goodness" of a fit (see Pevear and Schuette, this volume, for a modification which does); the fit is judged "by eye," guided by the knowledge of the investigator about the likely character of the sample.

ADDITIONAL CONSIDERATIONS

Peak Shape Modeling

Analysis of the shape of clay mineral peaks is rapidly becoming as important in the study of clay mineral diffraction patterns as peak position and intensity because of the wealth of information which can be obtained. Peak shape is controlled by many variables which can be divided into the following general categories: instrumental and sample controlled effects; strain effects; particle-size effects; defects effects; and mixed-layering effects. Instrumental and sample effects can be accounted for in NEWMOD© as discussed above. A more complete discussion of the relationship between the pure mineral peak shape and the instrumental signature is given by Jones and Malik (this volume). Strain effects can not presently be modeled with the NEWMOD© algorithm but can be effectively studied using the Reitveld method (Bish, this volume) or the Warren-Averbach method (Eberl and Blum, this volume). The effects of particle size, defects, and mixed-layering on clay mineral peak shapes are very important, and all are accessible through NEWMOD©.

Particle-size versus defect broadening. Figure 4 shows the effects of different particle size distributions on diffracted intensity. The top trace represents diffraction from an aggregate of particles each of which is 5 unit cells thick. The bottom trace represents diffraction from the more realistic situation of an aggregate of equal proportions of particles ranging in thickness from 3 to 14 unit cells. The ripples in the top pattern have been smoothed out in the bottom pattern because each particle size produces ripples in different places which merge into a smooth curve. The integrated areas of the two peaks are equal, and their breadths are proportional to the average unit cell thickness as described by the Scherrer equation. It is worth noting that the Scherrer equation is only valid in situations such as this where particle-size broadening is the major contributor to line broadening. Pure particle-size broadening as calculated using Eq. (8) will produce Gaussian peak shapes.

The presence of defects, such as stacking faults, within a particle will cause the particle itself to be an aggregate of coherent diffracting arrays, and will produce peaks which are more Lorentzian or Cauchy in form. X-ray diffraction peaks from clay minerals often exhibit both particle-size and defect broadening. The resulting shape is, therefore, a hybrid between the Lorentzian and Gaussian shape. This hybrid shape can be modeled by a number of peak shape

Guided Tour of NEWMOD©

functions including the Voight, pseudo-Voight, split Pearson, and Pearson *VII* functions (Howard and Preston, 1989).

An alternative to describing peak shapes by shape functions is to use the defect-broadening concept of Ergun (1970). If the distribution of interfaces (such particle boundaries or defects) which define the limits of diffracting arrays is random, the probability of occurrence for such an interface is given by exp($-n/\delta$) in which δ is the mean defect-free distance (Ergun, 1970), and Eq. (8) can be modified to

$$\Phi = \frac{N}{2} + \sum_{n=2}^{n=N-1}(N-n)\ e^{-n/\delta}\ \cos(4\pi R_s \sin\theta/\lambda). \quad (12)$$

The defect-free distance (δ) is expressed in terms of numbers of unit cells in a coherent X-ray scattering array. For large δ, the exponential term approaches 1. Domain sizes cover a range from $n = 2$ (LOWN in NEWMOD©) to $n = N-1$ (HIGHN in NEWMOD©) with the weighting of each domain size given by the probability of its occurrence. The upper limit (HIGHN) required to produce an appropriate peak shape is governed by the character of the mineral. If HIGHN is equal to or greater than 7δ, the peak shape is Lorentzian or Cauchy, indicating pure defect broadening. If HIGHN is smaller than 7δ, the peak shape becomes more Gaussian, indicating particle-size broadening. Reynolds (1985) recommends 5δ as a reasonable HIGHN for producing realistic peak shapes for many clay minerals, suggesting a combination of particle-size and defect broadening. Figure 5 shows the chlorite 004 calculated with several different particle size and defect distributions. Note how the peak shape changes from Gaussian to Cauchy with increasing HIGHN showing the relative effects of particle-size and defect broadening.

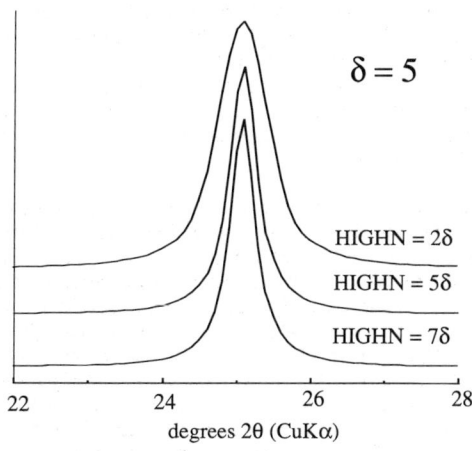

Figure 5. Chlorite 004 calculated with $\delta = 5$, and HIGHN ranging from $2*\delta$ to $7*\delta$.

Describing peak shape with defect broadening as given in Eq. (12) allows one to produce realistic peak shapes without having to resort to one of the various hybrid peak shape function listed above which rely on parameters which may be refinable but for which the physical significance may be unclear (see Bish, this volume). In some ways the approach in Eq. (12) is more satisfying than using the hybrid peak functions because peak shape is described in terms of measurable quantities such as defect-free distance and maximum number of unit cells, instead of deriving these quantities after the fact from coefficients in the peak shape function.

Mixed-layer broadening. Breadths of all diffraction peaks from a pure mineral, that is without instrumental effects or strain, are the same except for broadening with increasing angle which can be corrected for by multiplying by $\cos\theta$. Mixed-layer minerals have peak breadths which do not exhibit the same consistency. The peak breadth relationships are, however, systematic and can be used to quantify the amount of interstratification for random mixes. Méring (1949) astutely observed that diffraction peaks from a random interstratification of two different layer types occur between the locations of the peaks of the interstratified species, and that the breadths of the mixed-layer peaks are related to the distance separating the peaks which contribute to them. Figure 6 shows a calculated diffraction profile from a 50:50 random interstratification of chlorite and glycolated trioctahedral smectite. Also shown are the positions of the basal reflections from the chlorite and the smectite, demonstrating that, when the peaks are close together such as in the region between 5 and 6 °2θ, the resultant peak is sharp whereas when the peaks are widely separated such as between 16 and 19 °2θ, the resultant peak is very broad.

The principles of Méring have been quantified in the so-called "Q-rule (Moore and Reynolds, 1989) which can be applied as follows. For an interstratified system, the basal spacing of the minor component is divided by that of the major component to yield a factor. This factor is multiplied by the order of a diffraction peak. The deviation of the product from the nearest integer is called "Q," and describes the breadth of the peak in relation to the other peaks in the basal series. Values for Q range from 0, indicating no broadening, to 0.5 indicating maximum broadening. If peak breadth, corrected for diffraction angle, is plotted against Q, a linear relationship is observed, the slope of which is a function of the percentage of the minor component interstratified in the system. By creating a series of standard curves using NEWMOD© for different interstratifications, very small percentages of a minor component can be recognized in a sample which otherwise appears to be pure (see, for example, Walker and Murphy, 1993). Accomplishing this analysis, however, requires that the peak width be measured very carefully, that the instrumental signature be removed, and that no interferences are present.

Interstratification of small amounts of serpentine in chlorite have been described in several electron microscopic studies, however, it has rarely been identified by X-ray diffraction (Walker and Thompson, 1990; Reynolds et al., 1993) because of the near coincidence of the serpentine peaks with the even-order chlorite peaks. By inspection, interstratifications of serpentine and chlorite can be recognized because the odd basal reflections are broad whereas the even reflections are sharp. Using the Q-rule, a series of curves can be generated (Figure 7) by which the percentage of interstratified serpentine in chlorite can be estimated from the breadths of the diffraction peaks. One final *caveat* is that the analysis should usually include as many peaks as possible so that the slopes of the lines such as are shown in Fig. 7 are well-defined. Figure 7 was generated from the first 9 orders of the serpentine/chlorite interstratification. It is interesting to note also that extrapolating the curves to $Q = 0$ gives a measure of non-mixed-layer peak broadening, that is peak broadening due only to particle size or defect density.

Guided Tour of NEWMOD©

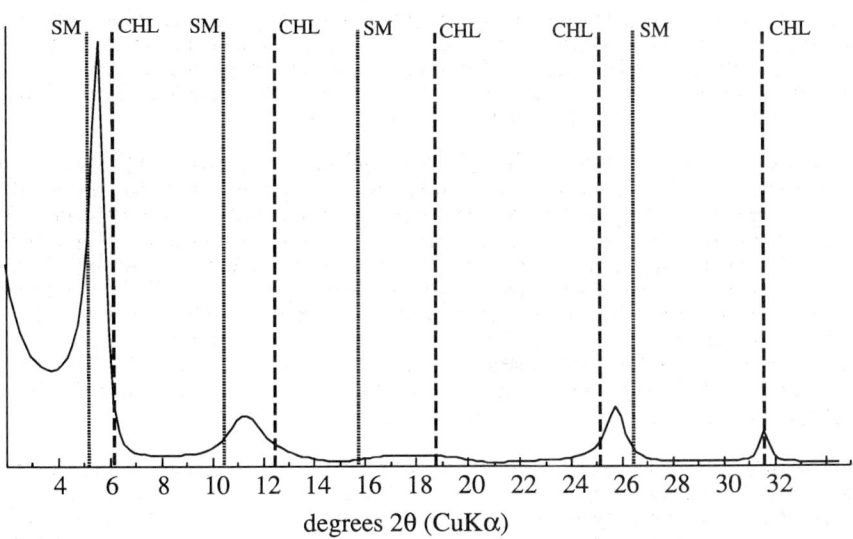

Figure 6. Calculated diffraction pattern from 50:50 mixed-layered chlorite:(glycol)smectite, $R = 0$, showing location of basal reflections of chlorite and trioctahedral smectite, to illustrate Méring's principles.

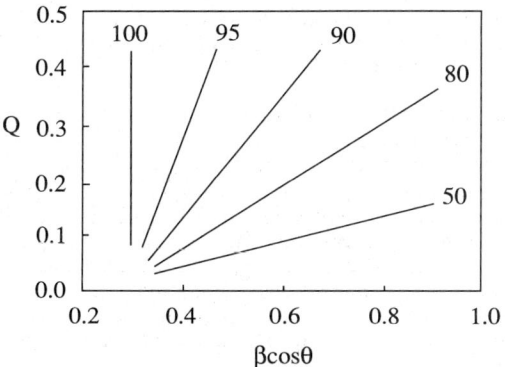

Figure 7. Series of curves relating Q and β for interstratified serpentine:chlorite, $R = 0$.

Modeling Phases Not Included In NEWMOD©

The discussion so far has been relevant to the most common applications of NEWMOD© that is, calculation of one-dimensional diffraction patterns from pure and interstratified examples of the basic clay types included in the program. However, there are times when clay types or other conditions which are not included in the program may need to be modeled. These situations commonly involve the inclusion of different atom or layer types.

Two approaches are possible to model scattering from atoms not included in the program. Because the scattering power of each plane of atoms is governed principally by the number of electrons in that plane, the first approach is to specify an amount of an atom type included in NEWMOD© which gives the desired number of electrons in the correct position. The actual size and number of substituting atoms is not important because all atoms are projected onto Z, however, if there is a large size discrepancy it may be necessary to adjust $d(001)$ to get realistic results. Alternatively, a scattering power curve for the atom type in question can be plotted from information in Lonsdale (1968), a 5-coefficient polynomial fit to the curve calculated, and the polynomial coefficients (Wright, 1973) added to the GTRANS segment of NEWMOD©. This approach, though not difficult, is more time-consuming than the former, but is justified for atom types which the user encounters regularly but are not included in NEWMOD©.

The procedure for specifying different layer types is similar to that just described for different atom types. After choosing the structure most closely resembling the mineral in question, the composition of the new layer type is approximated by inserting the desired composition in response to queries regarding K and Fe (i.e., talc would be TRIMICA with no K or Fe; Moore and Reynolds, 1989). The default layer thickness, $d(001)$, may also be adjusted. Tetrahedral substitution of Al^{3+} for Si^{4+} has little effect on the intensity because of the nearly identical scattering power of these two ions. Very different layer types, such as octahedral sheets like the phyllomanganates or brucite, can be modeled by modifying the Z-coordinates of atoms in a default structure to reflect the coordinates of atoms in the new structure. This approach is very time consuming because of the need to ensure that all references to the new structure within the program are correct, but it is again justified for a mineral type which will be modeled many times (see for example Holland and Walker, 1991). It is the only option for modeling layer types which are very different from the default layers included in NEWMOD©.

Compositional superstructures, in which ordering occurs between layers of the same mineral with different compositions, can also be modeled. Newer versions of NEWMOD© (e.g., NEWMOD.M2.1.4) automatically interstratify a layer with itself and set $R = 0$ if the proportion of a clay type is given as 1.0. To model a compositional superstructure in NEWMOD©, however, the proportion of the mineral type is given a value less than 1.0, and greater than or equal to 0.5, consistent with the character of the superstructure. The second layer is specified to be the same type as the first, and a Reichweite is specified. This approach allows the user to specify a different composition for each layer within the compositional superstructure. If the superstructure is not ordered, however, the results of this calculation will be indistinguishable from those of a mineral whose composition is the average of the two layers (Moore and Reynolds, 1989).

NEWMOD© can also produce calculated diffraction patterns from physical mixtures of two clay minerals using the utility MIXER©. Mixing is constrained in that patterns to be mixed must be calculated at the same angular increment, and the instrumental and sample conditions of the patterns

Guided Tour of NEWMOD©

to be mixed must be the same to produce physically realistic results. A possible exception to this last would be a situation in which there was good evidence that the two phases in question were oriented differently in the sample in which case their respective patterns could be calculated with different values for SIGMASTAR. The possibilities for performing quantitative analyses by such methods are discussed by Pevear and Schuette (his volume).

A version of NEWMOD© is available for calculating diffraction patterns from the interstratification of three layer types (NEWMOD3C©). In NEWMOD3C©, three components, A, B, and C are specified for which A and B may be related by any Reichweite. C is then substituted randomly for either A or B at the option of the user. It is beyond the scope of this chapter to discuss in detail the modeling of three-component systems. Also available is the program CLAYS© which treats simple, non-mixed layered clay minerals. It is very fast, and values of N up to 500 can be used to model the very sharp reflections observed for some kaolinites and chlorites.

CONCLUDING REMARKS

The NEWMOD© software series provides a powerful tool for modeling experimental XRD patterns of oriented clay minerals. The interface is "user-friendly," many important instrumental factors can be included in the model, default values can be easily changed to reflect laboratory-specific conditions, and the code has been optimized so that model runs can be made rapidly. There are several other uses of NEWMOD© which are not readily apparent but which are worth mentioning.

The standard and relatively simple format of data files produced by NEWMOD©, makes it easy for investigators to share not only calculated but experimental patterns as well. Because NEWMOD© data files are not compiled or encrypted in any way, they may be opened and read to determine their format. In order to create experimental data files in NEWMOD© format, all that is required is an ASCII data file from the X-ray diffractometer, and a simple program to put the ASCII data into NEWMOD© format. The data files can then be plotted using NEWMOD© plotting packages, and are also easily transmitted over computer networks. It could be that the NEWMOD© data files format will become a standard format for exchange of XRD analyses of clays.

In addition, NEWMOD© allows the user to produce diffraction profiles from pure systems, an option which is not always available to the experimentalist. In this way, an investigator can calculate standard patterns without interferences from secondary phases, and without having to worry about small percentages of interstratification which are commonly present in natural samples. Although this capability will never replace services such as the Clay Minerals Society "Source Clays," it does mean that the user has access to diffraction profiles of phases for which a natural material is not on hand. This idea can be extended to include materials which have not been described in nature. Limited only by one's imagination, and crystallochemical considerations, diffraction patterns from any kind of interstratification can be calculated, and criteria for recognizing previously unrecognized mixed-layered systems can be generated. Also, structural variations, such as inverted 7Å layers in chlorite, which may be theoretically possible but which have not been described, can be modeled. Some of these modeling possibilities require a thorough knowledge of the NEWMOD© code, but because the program is available in an uncompiled format, and because the theory is described in detail in numerous publication (see references cited herein), it is possible to make modifications to the code to suit the problem at hand.

ACKNOWLEDGMENTS

The author is indebted to R. C. Reynolds, Jr., for a critical reading of this chapter and for continual encouragement during repeated forays into the realm of crystal structure modeling of clays. I would also like to thank my students, especially K. L. Holland, M. P. Murphy, and L. S. Giovannini, whose questions and ideas have taken me deeper into NEWMOD© than I ever would have dreamed

REFERENCES CITED

Bethke, C. M. and Reynolds, R. C., Jr. (1986) Recursive method for determining frequency factors in interstratified clay diffraction calculations: *Clays & Clay Minerals* **34**, 224-226.

Ergun, S. (1970) X-ray scattering by very defective lattices: *Physical Review* **1**, 3371-3380.

Holland, K. L. and Walker, J. R. (1991) Crystal Structure of K-birnessite: *Programs and Abstracts, 28th Annual Meeting*, The Clay Minerals Society, Houston, TX, 75.

Howard, S.A. and Preston, K.D. (1989) Profile fitting of powder diffraction patterns: in *Modern Powder Diffraction*, D. L. Bish and J. E. Post, eds., *Reviews in Mineralogy* **20**, Mineralogical Society of America, Washington, D.C., 217-276.

Jadgozinski, H. (1949) Eindimensionale Fehlordnung in Kristallen und ihr Einfluss auf die Rontgeninterferenzen. I. Berechnung des Fehlordnungsgrades aus der Rontgenintensitaten: *Acta Crystallographica* **2**, 201-207.

James, R. W. (1965) *The Optical Principles of the Diffraction of X-rays*: The Crystalline State, Cornell University Press, Ithaca, 664 pp.

Lonsdale, K. (1968) *International Tables for X-ray Crystallography*: Kynoch Press, Birmingham, 202-207 pp.

Méring, J. (1949) L'Interference des Rayons X dans les systems a stratification desordonnees: *Acta Crystallographica* **2**, 371-377.

Moore, D. M. and Reynolds, R. C., Jr. (1989) *X-ray Diffraction and the Identification and Analysis of Clay Minerals*: Oxford University Press, Oxford, 332 pp.

Reynolds, R. C., Jr. (1965) An X-ray diffraction study of an ethylene glycol-montmorillonite complex: *American Mineralogist* **50**, 990-1001.

Reynolds, R. C., Jr. (1967) Interstratified clay systems: calculation of the total one-dimensional diffraction function: *American Mineralogist* **52**, 661-672.

Reynolds, R. C., Jr. (1968) The effect of particle size on apparent lattice spacings: *Acta Crystallographica* **A24**, 319-320.

Guided Tour of NEWMOD©

Reynolds, R. C., Jr. (1976) The Lorentz factor for basal reflections from micaceous minerals in oriented powder aggregates: *American Mineralogist* **61**, 484-491.

Reynolds, R. C., Jr. (1980) Interstratified clay minerals: in *Crystal Structures of Clay Minerals and their X-ray Identification*, G. W. Brindley and G. Brown, eds., Monograph **5**, Mineralogical Society, London, 249-304.

Reynolds, R. C., Jr. (1983) Calculation of absolute diffraction intensities for mixed-layered clays: *Clays & Clay Minerals* **31**, 233-234.

Reynolds, R. C., Jr. (1985) NEWMOD©, a Computer Program for the Calculation of One-Dimensional Diffraction Patterns of Mixed-Layered Clays: 8 Brook Road, Hanover, NH. 03755.

Reynolds, R. C., Jr. (1986) The Lorentz-polarization factor and preferred orientation in oriented clay aggregates: *Clays & Clay Minerals* **34**, 359-367.

Reynolds, R. C. (1989) Principle of powder diffraction: in *Modern Powder Diffraction*, D. L. Bish and J. E. Post, eds., *Reviews in Mineralogy* **20**, Mineralogical Society of America, Washington, D.C., 1-17.

Reynolds, R. C., Jr. (1989) Diffraction from small and disordered crystals: in *Modern Powder Diffraction*, D. L. Bish and J. E. Post, eds., *Reviews in Mineralogy* **20**, Mineralogical Society of America, Washington, D.C., 145-182.

Reynolds, R.C., Jr, DiStefano, M.P., and Lahann, R.W. (1993) Randomly interstratified serpentine/chlorite: its detection and quantification by powder X-ray diffraction methods: *Clays & Clay Minerals* **40**, 262-267.

Smith, D.K., Nichols, M.C., and Zolensky, M.E. (1983) POWD10, a Fortran IV Program for Calculating X-ray Diffraction Patterns: The Pennsylvania State University, University Park, PA.

Walker, J.R. and Murphy, M.P. (1993) Chloritic minerals in prehnite-pumpellyite facies metamorphic rocks of the Winterville Formation, Aroostook County, ME: in Day, H.W. and Schiffman, P.W., eds., *The Transition from Basalt to Metabasalt: Environments, Processes, and Petrogenesis*, GSA Special Paper, in press.

Walker, J.R. and Thompson, G.R. (1990) Structural variations in chlorite and illite in a diagenetic sequence from the Imperial Valley, California: *Clays & Clay Minerals* **38**, 315-321.

Wright, A. C. (1973) A compact representation for atomic scattering factors: *Clays & Clay Minerals* **21**, 489-490.

INVERTING THE NEWMOD© X-RAY DIFFRACTION FORWARD MODEL FOR CLAY MINERALS USING GENETIC ALGORITHMS

D. R. Pevear and J. F. Schuette

CONTENTS

Introduction	20
Genetic Algorithms	21
Overview	21
Genetic Algorithms 101	22
Pop Example	22
Step one: design a chromosome	22
Step two: build and evaluate a starting population	22
Step three: build the next generation and continue evolving	22
Genetic Algorithms 201	23
Contrasted to Other Techniques	25
MatchMod Revealed	25
Implementation Of MatchMod	25
Tour Of MatchMod with discussion	28
Choice of Manipulated Parameters: Input Dialog Boxes	28
The main dialog: mixed-layer phase	28
The discrete phase dialog	30
Choice of manipulated parameters	30
A MatchMod Run: The Run Window	31
Fitness Evaluation	33
Examples With Discussion	34
Cretaceous Shale	34
Coarse clay fraction	34
Medium clay fraction	34
Fine clay fraction	34
The Illite 001 Problem	36
Discussion And Conclusions	38
Factors affecting fit	38
Phase not in MatchMod	38
Multiple phases of the same broad mineral group	38
Simplistic mineral descriptions in the model	38
Instrument and sample parameters (site settings) incorrect	39
Quantitative analysis	39
Future Directions	39
Acknowledgements	39
References Cited	39

INVERTING THE NEWMOD© X-RAY DIFFRACTION FORWARD MODEL FOR CLAY MINERALS USING GENETIC ALGORITHMS

D. R. Pevear and J. F. Schuette

EXXON Production Research Co
P.O. Box 2189
Houston, TX 77252-2189

INTRODUCTION

We have developed a computer program, MatchMod, which inverts the NEWMOD© forward modeling program of Reynolds (1985a) and provides a "match" for a given experimental X-ray diffraction (XRD) pattern. As used here, a forward model is a set of equations, based on first principles, which can predict the state of a system from the variables which control that system. An inverse model, on the other hand, is one which estimates the control variables from some state of the system. Reynolds' series of tools (NEWMOD©, CLAYS©, MIXER©; Reynolds, 1985 a, b, c) have the potential for producing a quantitative clay mineral analysis based on the *00l* peaks of oriented clay mineral aggregates because they account for most instrumental factors, preferred orientation, sample length, a host of mineral compositional and structural variables as well as mass absorption coefficient and unit cell volume. However, in the practice of pattern-matching (inverse modeling) these tools are extremely time-consuming and tedious to apply, primarily because of the large number of variables involved in the trial-and-error procedure.

As Reynolds (1989a) notes: "The mathematics of the diffraction problem cannot be inverted [because the phases of scattered X-rays are not known], and the nature of the atomic arrangement in crystals can only be deduced by trial-and-error procedures." MatchMod is a special case of such a trial-and-error procedure in which the fit to an experimental pattern of four forward models (mixed-layer illite/smectite, discrete illite, discrete chlorite, and discrete kaolinite) is controlled by a genetic algorithm (Figure 1). Thus MatchMod is not a true inverse model, but rather an effective one. It does nothing that a skilled user could not do with Reynolds' tools; in fact, MatchMod is always able to incorporate user suggestions as it is running. Although MatchMod will take minutes or even hours for a run on a microcomputer, the amount of user time to find an acceptable match may be cut by a factor of 10 to 100.

Genetic algorithms (GA) can search a field of variables using a forward model and optimize variables in the model to fit a set of experimental data, in our case an XRD pattern. The standard NEWMOD© variables become "genes" on an individual "chromosome"; populations of genetic individuals crossover, reproduce and mutate. In each successive generation of individuals, only the "fittest," that is, the best fit to the experimental XRD pattern, are retained, and these then continue to crossover, reproduce and mutate to form the next generation. The mutation operation introduces randomness in order to prevent trapping in a "false maximum". After many generations the surviving individuals (sets of input variables) are so well adapted to their environment (an experimental XRD pattern) that we call it a "good fit".

Inverting Newmod© Using Genetic Algorithms

MatchMod superficially resembles the Rietveld method in that a forward model, based on the first principles of X-ray optics, is optimized to fit an experimental XRD pattern (Snyder and Bish, 1989; Post and Bish, 1989). However, Rietveld is more deterministic in that it uses a least-squares method, considers only Bragg reflections to which an instrumental profile is added, and uses three dimensional *hkl* data. The interference function, in which h, k, and l are treated as continuous variables (in units of 2θ; in our one-dimensional case using only l), which is required to model the small, disordered and mixed-layered crystals typical of clay minerals (Reynolds, 1989a), is not considered in Rietveld. The peak shapes, widths, etc. in MatchMod, as in NEWMOD©, are largely due to structural characteristics of the modeled mixed-layer minerals, rather than to the domain size, strain, and instrumental factors treated by the Rietveld method.

GENETIC ALGORITHMS

Overview

Many natural phenomena can be directly simulated by computer-based forward models; examples are found in macroeconomics, finite element analysis, and X-ray crystallography. Often, however, a user of these models really wants to "run the model in reverse" to discover the inputs that produce a particular result. The user may be trying to maximize some model output such as gross domestic product, or minimize some error such as the difference between a result predicted by the model and one observed in nature. The latter problem - finding the variables which control some result - is often referred to as the inverse modeling problem. MatchMod uses a GA to solve the problem of determining clay mineral composition based on XRD patterns. The model being "inverted" is Reynolds' NEWMOD© synthesis of XRD patterns for a given clay mineral.

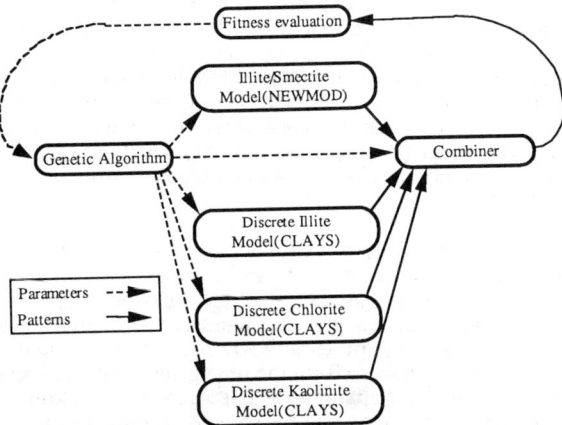

Figure 1. Structure of MatchMod showing information flow. MatchMod is controlled by the genetic algorithm, which uses the fit to an experimental pattern as a basis for sending new parameters to NEWMOD©, CLAYS© and the combiner.

Pevear and Schuette

Genetic Algorithms 101

Natural evolution is really an optimization process. New individuals evolve that share the "better" traits and omit the "worse" traits of their ancestors. Which traits persist and which die off are determined by the fitness, in some environment, of the individuals bearing those traits. Genetic algorithms (Holland, 1975; Goldberg, 1989) use fitness-based optimization to find near-optimal solutions to problem-specific objective functions. In MatchMod, the problem is to find the NEWMOD© input parameters that result in a particular observed XRD pattern. The technical details of GA design are well-described in Goldberg's (1989) book. A good introduction to the subject is Holland's (1992) article in Scientific American. In this section we use a simple "pop" example to give you a taste of how GAs work.

Pop Example You are the head of R&D at a leading beverage concern and are being pressured by management to develop a new hit soft drink to compete with the clear colas. There are many ways to tackle the problem, but the times demand a revolutionary strategy: you are going to *evolve* a new cola based on the principles of natural evolution.

Step one: design a chromosome

First, you must develop a representation of your cola analogous to the genome representation of living organisms. You decide that an array of ingredient proportions can capture all colas of interest and design the following *chromosome*.

%Carbonated Water	%Sugar	%Citric Acid	%Potassium Sorbate	%Flavors	%Caffeine

To simplify the process, you have already decided on a set of natural flavors to use, so the only question remaining is how much of each ingredient you want to include.

Step two: build and evaluate a starting population

Natural evolution requires populations of individuals that can breed and compete with each other, so you generate 100 random chromosomes to represent your initial population. As described above, natural evolution is a fitness-based optimization technique, so you need a way to figure out which individuals (cola formulations) are more fit than others. You decide that a simple taste test lets your potential customers rate the individuals and gives you the basis for making a fitness assignment to each individual.

Step three: build the next generation and continue evolving

You now have a generation with a fitness assigned to each individual and you need to breed a new generation. In general, there are three ways to put an individual from the current generation into the next generation: you can simply copy it (technically, this operation is called *asexual reproduction*), you can copy it including some random change (*mutation*), or you can create a new individual by cutting and pasting (Figure 2) sections of two previous individuals (*genetic recombination*). The latter two options are how you discover new possibilities. You randomly select individuals on which to operate, but not with uniform probability. Remembering that evolution is fitness-based, you make the selection probability of each individual proportional to its

Inverting Newmod© Using Genetic Algorithms

fitness.

Conceptually, by using a fitness-based selection procedure to pick individuals for the reproduction, mutation, and genetic recombination operators, you are searching for better solutions by incorporating bits and pieces of prior superior solutions. After building a new generation, you have another taste test to assign fitnesses to those individuals, generate another new population, and iterate until you are satisfied with the results (perhaps when your best formula beats the competitors' formula in a taste test).

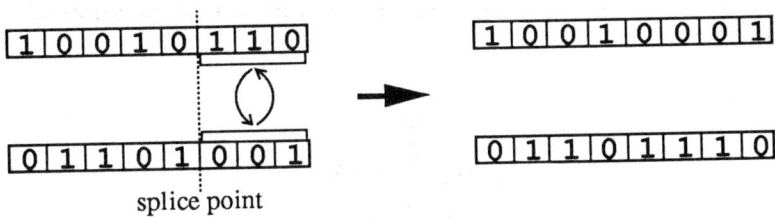

Figure 2. Genetic recombinaton is simulated by the crossover operator schematically shown here. The genetic algorithm uses binary "chromosomes" representing model parameters to generate new individuals for consideration. The left half of the figure shows a randomly picked splice point and the swapping action of the crossover operator. The right half of the figure shows the resulting new individuals.

Genetic Algorithms 201

Genetic algorithms are inspired by biological concepts such as genetic recombination, survival-of-the-fittest, and mutation and have been successfully used for a wide range of applications such as function optimization, engineering design optimization, and scheduling (Goldberg, 1989). Because they can efficiently search non-linear solution spaces, they are well-suited to solving inverse modeling problems. There are many design issues that must be considered in using a GA to solve a particular problem, but such discussions are detailed in (Goldberg, 1989). The features common to almost all GAs are briefly discussed here.

A GA manipulates *populations* of *individuals* to find the individual with the highest *fitness*. In the case of an inverse modeling problem, each individual is a set of input parameters to the forward model. The fitness of an individual is the degree to which the output corresponding to that individual matches the desired output: therefore the evaluation of the fitness of an individual corresponds to an execution of the forward model. The GA evolves populations of more highly fit individuals by using the operations of *crossover*, *reproduction*, and *mutation* (Figure 3).

Pevear and Schuette

Some crucial facts must be noted before summarizing how the operators work. The only way an individual can "survive" into future populations is to be operated on by one of the genetic operators. Furthermore, the operands of each operator are individuals picked with a probability in proportion to their fitness. These two facts explain how the GA concentrates its search on the better individuals. Finally, the individuals are actually encoded representations (usually binary) of the original parameters. The operators are purely syntactic operations. In other words, the genetic search in MatchMod is in no way burdened with knowledge about clay minerals or XRD patterns.

The *crossover* operation randomly selects two individuals (with probability in proportion to their fitness), picks a random splice point, swaps the ends of the two chosen individuals, and places the offspring in the next generation (Figure 2). This corresponds to biological genetic recombination and allows the search to consider new (possibly fitter) possibilities.

The *reproduction* operator randomly selects one individual (with probability in proportion to its fitness) and copies it into the next generation. This operator corresponds to the biological concept of survival-of-the-fittest.

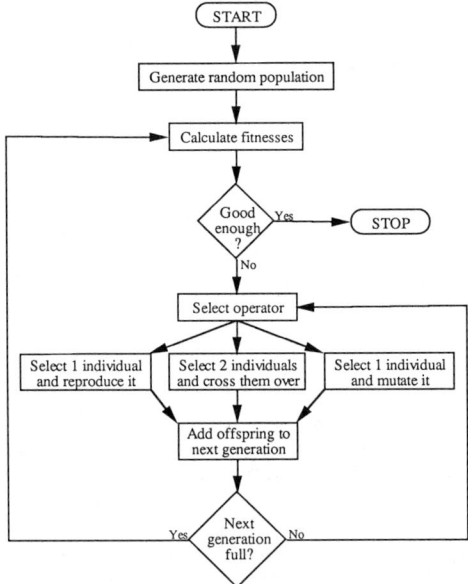

Figure 3. Operation of the Genetic Algorithm in MatchMod. The "Calculate fitness" box includes the calculation of modeled XRD patterns from a set of input parameters, and evaluation of their fit to an experimental pattern. The initial population of individuals (individual sets of parameters) is random (top); thereafter the fit of the population is continuously improved (loop) by the three operators (reproduction, crossover, mutation) until a stop condition is reached.

Inverting Newmod© Using Genetic Algorithms

The *mutation* operator randomly selects one individual (with probability in proportion to its fitness), flips a random bit, and copies it into the next generation. Mutation is actually a relatively minor contributor to genetic search (as it is in nature).

The GA uses these operators to build more fit populations from previous ones and stops when some predetermined condition is met. That stopping condition may be the discovery of an individual whose fitness surpasses some minimum acceptable value or the completion of some constant number of population evolution cycles. When to stop a genetic search is one of the design issues alluded to at the beginning of this section; other design choices and typical values are summarized in Table 1.

Design choice	Typical values
Population size	50–500
Crossover probability	0.60
Reproduction probability	0.39
Mutation probability	0.01

Table 1. Some typical design choices

Contrasted to Other Techniques

The Monte Carlo and simulated annealing search methods also rely heavily on stochastic decisions. Monte Carlo methods don't, however, benefit from a fitness guided search. Simulated annealing shares many features with GAs by concentrating on improving fitness by random processes, but simulated annealing lacks both the large population sizes that allow progress on many fronts and the concept of crossover which allows non-local improvement. The major difference between GAs and hill-climbing techniques is that in a genetic search there is no need for derivative information about the space being searched. All that is required is to be able to determine a fitness for each case.

MatchMod Revealed

MatchMod's design is detailed below, but the only real differences between it and the pop example are that MatchMod's chromosome represents input parameters to NEWMOD© and the fitness is the degree to which the NEWMOD© output resulting from a particular individual (set of input parameters) matches an observed XRD pattern.

IMPLEMENTATION OF MATCHMOD

The widely used NEWMOD© (Reynolds, 1985a) calculates XRD patterns for the 00*l* reflections of a variety of mixed-layer clay minerals. Lesser known is CLAYS© (Reynolds, 1985b), which is similar to NEWMOD© but models discrete clay mineral phases; its main advantage is that it runs much faster than NEWMOD© (which can also model discrete phases) because it doesn't do the mixed-layer statistical calculations. These two programs were rewritten to run in C on the Macintosh, and form the core of the development version of MatchMod described here. The C version is faster than the original TrueBasic, and permits greater cross-platform transportability, particularly to workstations and multiprocessor computers. The reader is referred to Reynolds (1980, 1985a, b, 1989b) for details of the theory and operation of NEWMOD© and CLAYS©.

The individuals (chromosomes) manipulated by the GA represent various parameters (genes) to five processes (Figure 1). The first (NEWMOD©) synthesizes a mixed-layer illite/2-glycol smectite (I/S) XRD pattern, the second (CLAYS©) the discrete illite pattern, the third the discrete (tri-tri) chlorite pattern, the fourth the discrete kaolinite pattern, and the fifth combines (like Reynolds' MIXER©) the four XRD patterns to produce a consolidated result to compare with the experimental pattern. The 15 parameters manipulated by the GA are listed in Table 2 with the number of possible values for each. For example, K (potassium) is varied between 0 and 1 (per $O_{10}(OH)_2$) in 16 steps, Reichweite between 0 and 3 in 32 steps and so on. These steps result from the binary code used in the GA: 16 steps require 4 bits, 32 steps 5 bits, etc. The search space for this problem contains about 4×10^{24} possibilities. Rough calculations show that if we had been trying one new combination every microsecond since the big bang, we would at present have covered only 10% of the search space.

Parameter	Used to	Possible values
Sigma Star (σ*)	Generate all XRD patterns	32
Proportion of I in I/S	Generate I/S XRD pattern	128
Fe value for I in I/S	"	32
K value of I in I/S	"	16
Fe value for S in I/S	"	32
Reichweite for I/S	"	32
Fe value for Illite	Generate I XRD pattern	32
K value for Illite	"	16
nHyd value for Chlorite	Generate C XRD pattern	16
Fe1(sil.) value for Chlorite	"	32
Fe2(hyd.) value for Chlorite	"	32
Proportion of I/S in sample	Consolidate XRD patterns	128
Proportion of Illite in sample	"	128
Proportion of Kaol in sample	"	128
Proportion of Chl. in sample	"	128

Table 2. Genetically manipulated parameters

Inverting Newmod© Using Genetic Algorithms

The 15 parameters in Table 2 fall into several groups: Sigma Star (σ^*), a measure of preferred orientation in the sample, is assumed identical for all four minerals in a given sample. The next 10 parameters describe various compositional aspects of the four individual modeled clay species. The first 11 parameters are identical to their counterparts in NEWMOD©. The last four are the decimal fractions of each mineral used by the "Combiner" (MIXER©) in Figure 1; these will always sum to unity and constitute the "quantitative mineral analysis". The intensities generated by NEWMOD© and CLAYS© are adjusted by the unit cell volume (Reynolds, 1983), so simple addition of proportioned patterns should be equivalent to the weight proportion (or %) of the mineral in the sample, assuming a transparent matrix. The addition of 4 proportioned calculated patterns to produce a mixture is shown in Figure 4; it is such a mixture that is compared with an experimental pattern during the "Fitness Evaluation" step shown in Figure 1.

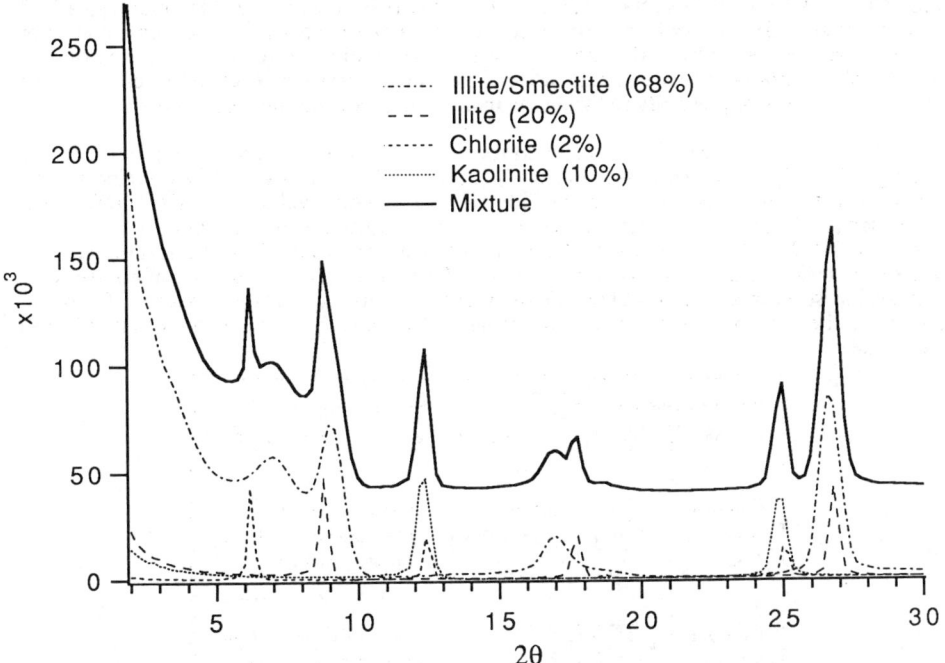

Figure 4. Mixture produced by addition of four modeled patterns, showing operation of the Combiner of Figure 1. The intensities of each model are multiplied by the decimal fraction of the phase in the Mixture (Reynolds,1985c). The Mixture pattern has been raised in the figure for clear presentation. Specific phase parameters are:
   I/S --Illite Layers 0.8, Illite Fe 0.4, Illite K 0.7, Smec. Fe 0.2, Reichweite 1.0, N 5-10, Sigmastar 3, Exchange Cation Na.
   Illite --Fe 0.0, K 0.9, N 10-25, Sigmastar 3.
   Chlorite -- Tri-Trichlorite Fe1(sil.) 2.6, Fe2 (hyd.) 0.5, Hyd 0.8, N 10-25, Sigmastar 3.
   Kaolinite --N 10-25, Sigmastar 3.

TOUR OF MATCHMOD WITH DISCUSSION

Choice of Manipulated Parameters: Input Dialog Boxes

The main dialog: mixed-layer phase. NEWMOD© includes many parameters not manipulated by the GA (not in Table 2); a critical aspect of MatchMod is the choice of parameters to manipulate, as discussed below. The main MatchMod dialog (Figure 5) displays additional user-adjusted parameters (all boxed items can be changed); most of these are saved in a preferences file and will remain for successive runs unless changed by the user. The **Maximum** number of **Generations** can be specified (or left indefinite). Each generation contains 50 individuals (50 patterns to compare with experimental), so 50 generations is 2,500 patterns. The 2θ range of the experimental pattern is initially shown in Figure 5. This range can be reduced to model only part of the pattern, which will greatly speed up run time. The **Increment** is the final 2θ step of the modeled pattern. The model is initially run in huge steps (five times the set Increment, so a 0.2° increment starts at 1 °2θ) so it will be fast. When the set **Threshold** value (between 0 and 1) of the fitness parameter is reached the step of the next generation shifts to four times the set **Increment** and the process repeats in regular steps until the set value of **Increment** in the dialog is reached. The process of gradually reducing the increment greatly diminishes run time.

Except for **Sigma Star**, which applies to all four mineral models, the remaining items on the main dialog refer only to the I/S phase, and are identical to their NEWMOD© equivalents. The **Coarse, Medium, Fine** items refer to the "lowN" and "highN" values in NEWMOD©, but allow the user to chose with a button between three possibilities which roughly relate, in our experience, to 2-0.2, 0.2-0.02, and <0.02 μm particle size-fractions. These fractions refer to equivalent spherical diameter calculated using Stoke's Law. The real particles are, of course, very thin, flat, and of somewhat different size. TEM measurements suggest the N values shown are reasonable, but they can be changed to any other values by the user. The exchange cation is also selected by user.

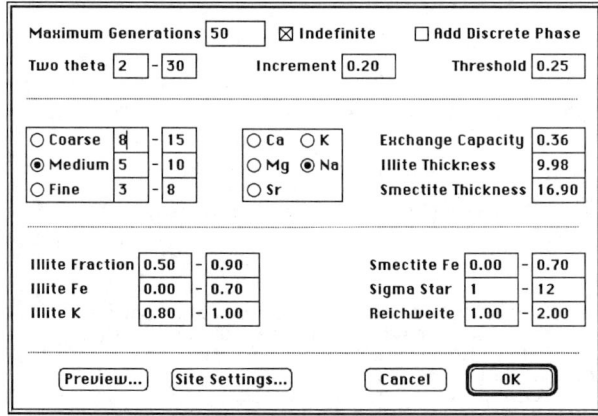

Figure 5. Main MatchMod Dialog, including I/S phase settings; details are discussed in text.

Inverting Newmod© Using Genetic Algorithms

The lower main dialog parameters are those manipulated by the GA. Note that many can be specified as ranges by the user. Thus the search space can be constrained to specified ranges if the user thinks he knows something about the sample; for example, K <0.5 in illite is likely unrealistic. If a model fit by MatchMod displays parameters at the limits of a range set by user, then it is possible that too narrow a constraint range was chosen; the model should be rerun with a wider range. To help the user specify the dialog items, the **Preview** button at the bottom opens a movable window displaying the experimental pattern to be matched (Figure 6). The **Site Settings** button at the bottom opens a dialog (Figure 7) familiar to NEWMOD© users in which instrument-specific parameters and the mass absorption coefficient of the sample are set. The mass absorption coefficient is mainly required where absolute intensities are compared between experimental and calculated patterns The specific parameters shown in Figure 7 apply to all experimental and calculated patterns discussed here.

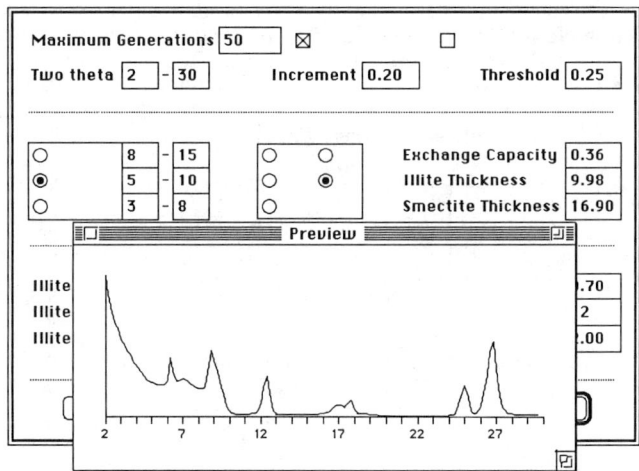

Figure 6. Preview window permits inspection of experimental pattern for inspiration during selection of user-adjusted input parameters.

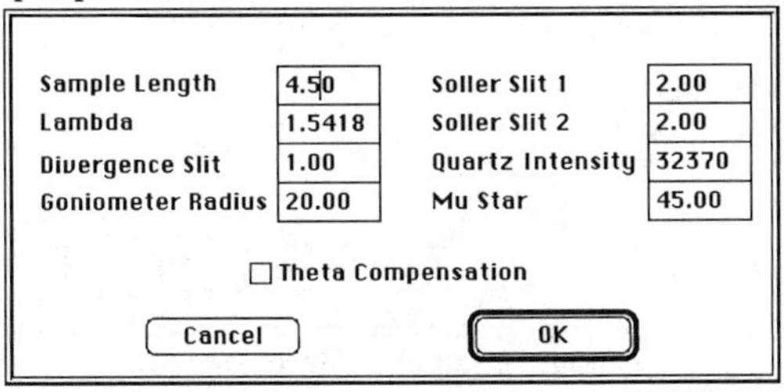

Figure 7. MatchMod site settings dialog, which resembles that in NEWMOD©

The discrete phase dialog. The **Add Discrete Phase** button at the top of the main dialog opens the discrete phase dialog (Figure 8) where the user selects the lowN-highN values (left side) and the ranges over which the GA will search (right side). If this button is not activated, then MatchMod only considers an I/S with 2 glycol layers. This dialog also allows the user to constrain the **Fraction** of each of the four phases (0-0 means that phase will not be considered). **Fe1** is the iron in the silicate octahedral site and **Fe2** the hydroxide octahedral site in tri-trichlorite. The **Thickness** values are the $d(001)$ of the respective unit cells (in Å), and are not manipulated by the GA.

Choice of manipulated parameters. The choice of parameters to be manipulated by the GA is critical to MatchMod's operation and is not entirely straightforward. Certainly we do not want to manipulate fixed parameters that we know, such as those of the settings dialog (Figure 7). Generally, we manipulate parameters which exert a significant control on the XRD patterns, most of which are structural and compositional properties of the sample, that is, things we want to know. Sigma Star, although not exactly in this category, must be considered if the modeled pattern extends over a large 2θ range because of its effect on the angular dependence of the Lorentz factor as it is treated in NEWMOD© (Reynolds, 1986). Since slit configuration, sample length, sample thickness, σ* and mass absorption coefficient (mu star, μ*) all affect intensity systematically with 2θ, the GA would likely be unable to differentiate them, so all but σ* are fixed.

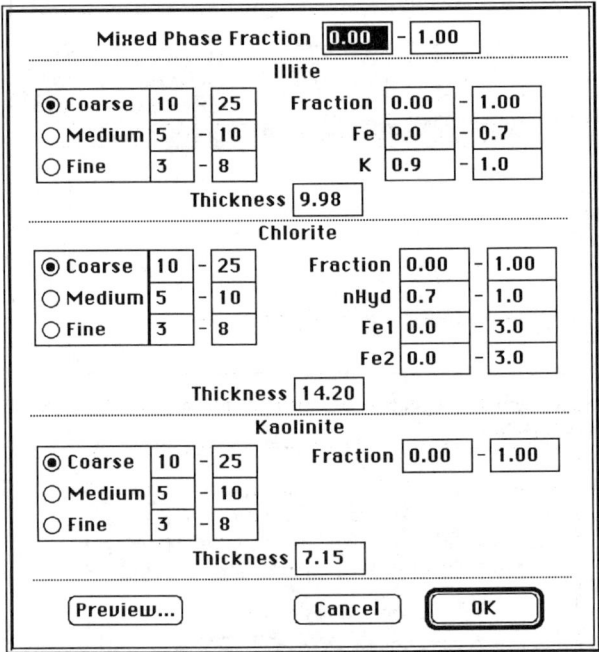

Figure 8. MatchMod discrete phase dialog; details explained in text.

Inverting Newmod© Using Genetic Algorithms

Two important NEWMOD© parameters, $d(001)$ and N are not presently manipulated by the GA because, for I/S, their effects on the XRD pattern cannot be cleanly separated from those of other parameters. This is particularly the case for $d(001)$. Since we "know" the illite $d(001)$ of a sample to a much greater degree than we "know" Reichweite, the former is fixed and the latter is manipulated. Future versions of MatchMod will likely include N (and defect broadening) as a manipulated parameter. The **OK** button on the main dialog (Figure 5) activates the MatchMod run window. This window permits the user to view the fit and interact with the GA as it is running.

A MatchMod Run: The Run Window

An example of a run window is shown in Figure 9, above which are displayed successively earlier versions of the same model run (broken line) compared with the experimental pattern (solid line). In this example we are using the mixture from Figure 4 (made using NEWMOD©) as an experimental pattern, since we know all of its parameters and can, therefore, see if MatchMod is able to extract them. The top pattern (a) is from the first generation and is a random arrangement of the parameters in Table 2; note it is in increments of 1 °2θ. The bottom pattern shows the fit after 12 minutes; the model values for the 15 parameters manipulated by the GA are shown in the first two lines. The first line is for the I/S (2-glycol) phase, and the number (0.65) under the "I/S" is the proportion (65%) of I/S in the model (compared to the actual value of 68%, see Figure 4). The second line shows values for the three discrete phases. The values for all the input parameters (manipulated and fixed) for the "experimental" pattern are in Figure 4; comparison with the model results in Figure 9 shows that the GA in MatchMod did a great job of inverting the forward model; it even found 2% chlorite (in the presence of 10% kaolinite) and the asymmetry of chlorite Fe distribution. The failure to find an exact fit is likely due to the fact that the parameter values shown are roundoffs of the powers-of-2 step series in Table 2. The user may pause the run at any time with the **Pause/Continue** button to adjust the manipulated parameters. If user's suggestion is a better fit than that displayed, the new fit will be displayed. If not, the suggestion is still saved in the population to compete with other individuals for a place in the next generation. Some pieces of the suggestion may survive, others may perish.

The bottom line in Figure 9 shows information about the GA: there have been 10 **Generations** (50 individuals each) and therefore 500 patterns were generated and compared with the experimental pattern. The **Countdown** starts at 50 and continuously drops as the 50 individuals (sets of parameters) in a generation are used to generate patterns to be compared with the experimental pattern. The current best fit is the one displayed as a patterned line. The **Resolution** (in °2θ) starts at 1° and steps down to the increment value set in the main dialog; current resolution of the model is displayed in the box, but the experimental pattern is always shown at the value set in the main dialog. The **Fitness**, a number between 0 and 1 (1=perfect fit) is discussed below.

The box on the lower right shows the value for a derived result, **Id** (proportion of discrete illite of the total illite, discrete and in I/S) which is used in our laboratory for an isotope dating routine (Pevear, 1992). The **best** is the current best fit; statistics for all the Id values (after the first, random generation) that were displayed during the run are given, and at the bottom is a 0.0 to 1.0 bar showing all the previous "best fit" Id values and the current best one. The information in this box is useful for evaluating the range of Id values that constitute a reasonable fit, and for estimating the precision of the Id "determination".

Pevear and Schuette

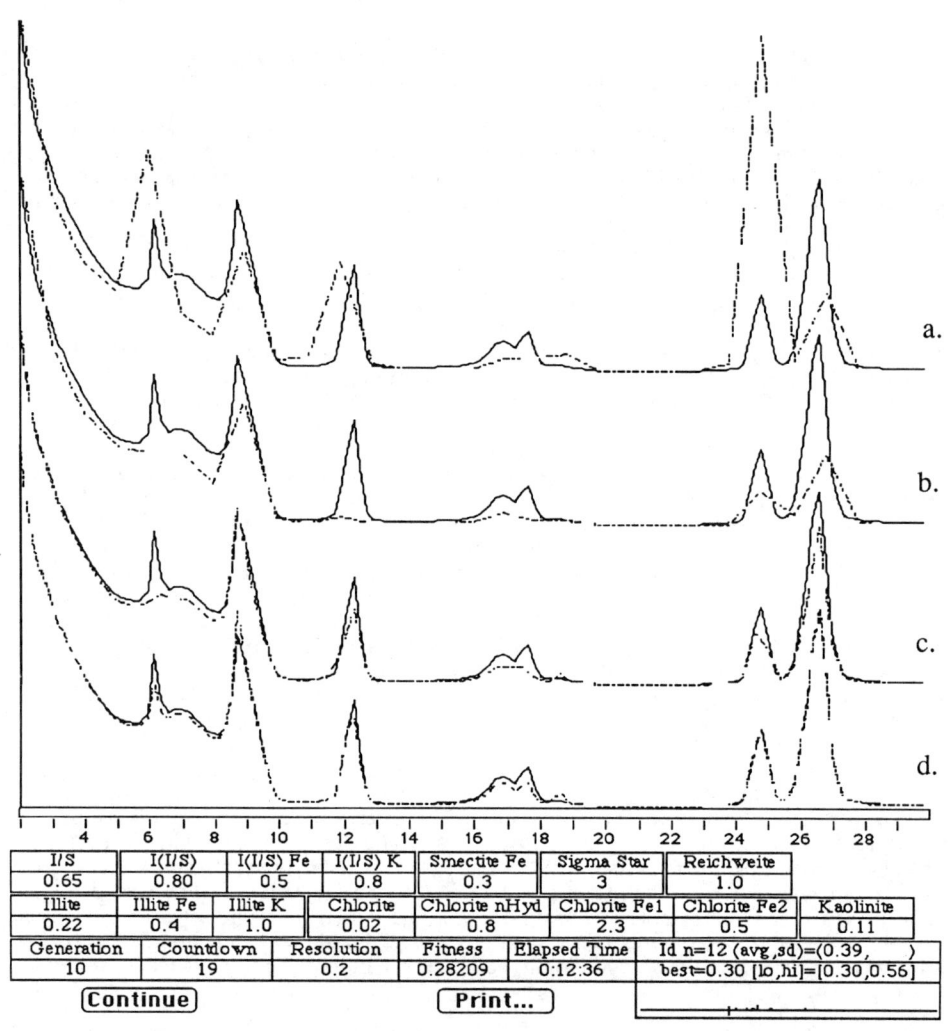

Figure 9. MatchMod run window (bottom, d) to which have been added (a, b, c) the patterns from successive previous (less perfect) states of the same run. Displayed values (explained in text) are for the best match. Solid line is the experimental pattern, broken line is the MatchMod calculated match. In this example the pattern being matched was also calculated, and is the same as that in Figure 4. The site parameters for this and all other figures are those of Figure 7.

Inverting Newmod© Using Genetic Algorithms

Fitness Evaluation

The quality of the result produced by an individual (set of parameters) is the basis for determining the fitness of the individual and therefore determining the probability that the individual (or pieces of it) will survive in future generations. The fitness values in MatchMod are in the range (0.0, 1.0) inclusive. A fitness of 1.0 implies a perfect match between the calculated pattern and the observed experimental pattern. However, as we shall see, a fitness of only 0.1 is generally, by inspection, an acceptable match.

The fitness of an individual in MatchMod is calculated by comparing the individual's calculated XRD pattern with the observed pattern. The fitness f is defined as

$$f = \frac{1}{1 + err}$$

where $0 \leq err \leq \infty$ and $0 \leq f \leq 1$, and err is a sum of squares error measured at each calculated point in the pattern:

$$err = \sum_{i=1}^{n} (calc(2\theta_i) - obs(2\theta_i))^2$$

where N is the number of 2θ points actually calculated by the forward models, $calc(2\theta_i)$ is the calculated intensity at a particular point, and $obs(2\theta_i)$ is the observed intensity at a particular point. Observed and calculated intensities are normalized to fall in the range (0.0, 10.0), inclusive, to produce fitness values in a range helpful to the user. The fact that the absolute magnitude of the fitness is insignificant is described below.

Along with population size, mutation probability, etc., the choice of a fitness measure is one of the design choices made by the algorithm author. This fitness measure was chosen because it is simple and because the squared error tends to emphasize areas of major disagreement between the calculated and observed patterns and "pay less attention to" areas of general agreement.

One consequence of the fitness measure and the fact that resolution increases with run time is that the absolute magnitude of the fitness may actually decrease from generation to generation. Therefore, for two matches with different resolutions but the same "visual quality," the fitness will be lower for the higher resolution match because it has more error terms. This apparent decrease in fitness with increasing resolution has only a psychological impact, however. The resolution only changes between generations, so all individuals competing with a given individual have the same resolution and therefore the same "fitness scale." Furthermore, remember that genetic search is guided only by the relative fitnesses of the individuals in the population, so the search will progress regardless of the absolute fitness values.

Genetic search is fitness-driven, but stochastic, so it is hard to predict how long a search must run before an acceptable match is found. MatchMod can be set to stop after a preset number of generations, but it is almost always run until the user is "satisfied" with the match. This could take minutes or hours, but MatchMod is always quicker than the alternative of the expert searching for matches by manually tweaking parameters to the forward model. Performance issues such as time and quality of the results are discussed in the next section. We generally consider a search done (for a pattern with a 30° range) when the final resolution is reached and the fitness parameter is > 0.1, or if there is no improvement after an hour or more. But there are exceptions, and the opinion of a skilled user is often helpful.

EXAMPLES WITH DISCUSSION

Cretaceous Shale

Coarse clay fraction. An example (Figure 10) from the 2-0.2 μm fraction of a Cretaceous shale reveals additional features and problems of MatchMod. The calculated pattern of Figure 4 was configured to resemble this real pattern. The open bar above the 2θ scale indicates the model range set in the main dialog; the entire experimental pattern is shown. The diagonal pattern areas within this bar, made by dragging the mouse, eliminate specific ranges from being used in the fitness evaluation, perhaps because quartz or other minerals not modeled are present, or because specific portions of the pattern are more important.

For anyone who has tried to make a synthetic fit "by hand" using NEWMOD©, CLAYS© and MIXER©, the fit in Figure 10 looks rather impressive. The compositional parameters for the I/S are nearly identical to those of a nearby bentonite which is monomineralic. However, note that the fit parameter never got to the 0.25 threshold in order to reach the set resolution of 0.2 - this is a typical problem with patterns covering a 30° range - the threshold (Figure 5) should be lowered. The kaolinite-chlorite doublet near 25° is not perfect either. The 12 hour run time was probably not necessary to get a good fit, as the cluster of nearly identical Id values in the lower right box suggests, but reflects the user going home for the night.

Medium clay fraction. The 0.2-0.02 μm fraction of the same sample (Figure 11), as expected, shows more I/S and less of the discrete phases. The various compositional parameters are in reasonable agreement with those of the coarse fraction. The low σ* value of 1 is cause for concern. Reynolds (1989c) observes, based on actual measurements of preferred orientation, that σ* values as low as 3 are rare and indicate unusually good alignment of platy crystals in the oriented sample. Our samples were prepared by the suction and transfer method, not known to produce the very best orientation (Moore and Reynolds, 1989). In fact, the low σ* value in Figure 11 (and 12) is incorrect; MatchMod is trying to compensate for a variable not under the control of the program, namely, sample thickness as discussed in the next section.

Fine clay fraction. The <0.02 μm fraction of the same sample is shown in Figure 12; the long run time again reflects a sleeping user, but does show that improvements, generally incremental, continue even after 16 hours (the **Elapsed Time** value is the time, since start, that the last, displayed, improvement was made). As expected, this sample contains more I/S than the two coarser fractions. The other parameters are consistent with those of the coarser fractions. The Id box in the lower right shows a bimodal distribution; early models had much more discrete illite than the final models, which produced better fits. Note the misfit in the chlorite peak near 12°; there is only 2% chlorite in the model, choice of 1% produces a larger misfit than the one shown, and 1.5% is not possible due to the step size the GA uses (Table 2).

Inverting Newmod© Using Genetic Algorithms

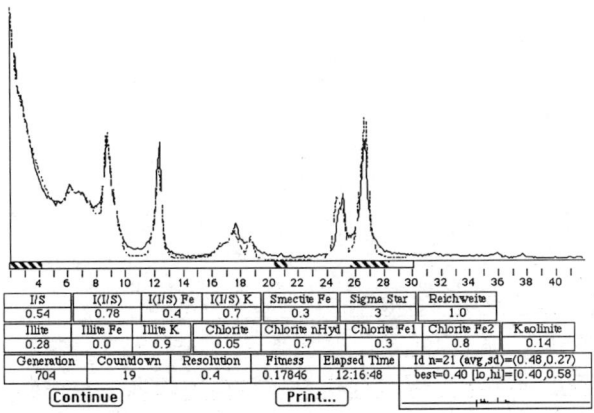

Figure 10. MatchMod run of the 2-0.2 μm (coarse clay) fraction of a Cretaceous shale; details described in text. Solid line is the experimental pattern, broken line is the MatchMod calculated match. The highN-lowN values shown in Figures 4 and 7 for the respective size fractions apply to this and the succeeding 2 figures.

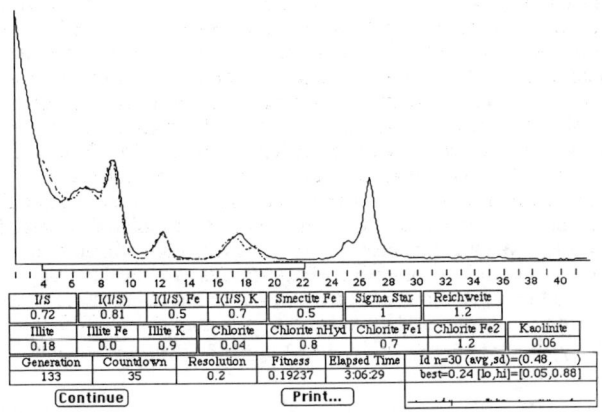

Figure 11. MatchMod run of the 0.2-0.02 μm (medium clay) fraction of the shale shown in Figure 10; details described in text. Solid line is the experimental pattern, broken line is the MatchMod calculated match.

Pevear and Schuette

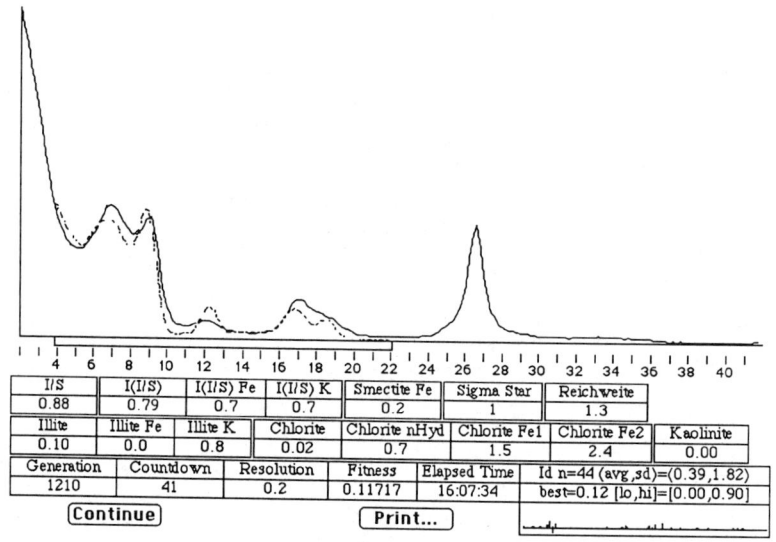

Figure 12. MatchMod run of the < 0.02 μm (fine clay) fraction of the shale in Figures 9 and 10; details described in text. Solid line is the experimental pattern, broken line is the MatchMod calculated match.

The samples in Figures 11 and 12 are not effectively "infinitely thick" so the peaks at higher 2θ angles are diminished in intensity with respect to those at lower angles, compared to a thick sample (see Reynolds, 1989c; Moore and Reynolds, 1989; Bish and Reynolds, 1989). The treatment of the Lorentz-Polarization factor in NEWMOD© introduces an angular dependency between σ^* and intensity (Reynolds, 1986) with which the GA tries to compensate for the thin sample. Of course the best (only?) way around this is to have an adequate sample. Although an expression for sample thickness (really mass/unit area) could easily be added to MatchMod (Bish and Reynolds, 1989), it would have to be a fixed (known) parameter just like sample length; if it were a manipulated variable the GA would probably be unable to distinguish its effects from those of σ^*. It is not clear that "effective thickness" could be easily and routinely measured, although Reynolds (1989c) offers some suggestions. We are currently working on using the absolute intensity of the silver XRD peak from samples mounted on silver membrane filters as a measure of sample thickness. A lazy way out, used here, is to consider a more limited angular range, over which the effects of sample thickness are not so severe. However, every degree of the XRD pattern contains information; using excessively short segments of the pattern in MatchMod gives good-looking matches, but not a unique solution.

The Illite 001 Problem

A final example illustrates a limitation which appears to reside in NEWMOD©, or more specifically, in our understanding of the nature of illite. Figure 13 shows an attempt to fit a pure

Inverting Newmod© Using Genetic Algorithms

illite experimental pattern by setting all the other mineral fractions to zero. The illite is a <0.02 fraction from a shale and shows almost no changes with glycol. In order to model the peak width a 2-6 lowN-highN range was used. The experimental (real) illite 001 is at 8.8° as expected, but the modeled 001 is shifted by about 0.5° toward lower angles. The shift is due to the rapidly-rising structure factor (G^2 or F^2) across the peak (from right to left) which enhances the low angle side and causes the shift. The Lorentz-Polarization factor also contributes to the observed peak displacement. Such a shift was reported for experimental patterns of hydrothermally synthesized mica for which TEM measurements of N were in the same 2-6 range as used above (Klimentidis, et al.,1991).

The question is not what's wrong with NEWMOD©, but why do we not see this shift in fine (broad-peaked) fractions of natural illites, which all seem to be at 8.8°? It can't be that $d(001)$ is wrong in the model, for it must be reduced from 9.98 Å to 9.6 Å in order for the 001 reflection to be at 8.8° and match the experimental pattern, and then the other peaks don't fit. For MatchMod, the illite 001 problem applies only to samples with highN < 10, where it is trying to fit an experimental peak at 8.8° with a modeled one at maybe 8.4°-8.7°. It is seldom a major difficulty; typically, as in Figure 13, the illite peak is matched, but is slightly displaced with respect to the experimental XRD pattern. But it is disquieting, and makes one feel a certain lack of confidence in the overall results, and concern that we really do not know the structure of illite in detail. The difficulty really is an illite problem, rather than a NEWMOD© problem, because it is not observed with other minerals, such as chlorite.

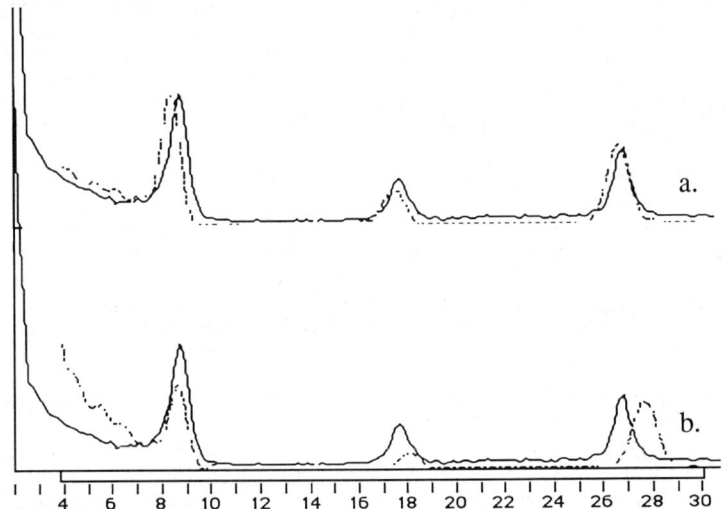

Figure 13. MatchMod run constrained to model illite only; N = 3-7. Upper pattern (a) shows run with $d(001)$= 9.98 Å (note misfit of 001 peak). Lower pattern shows run with $d(001)$= 9.6 Å; 001 peak now fits, but other peaks are offset. Solid line is the experimental pattern, broken line is the MatchMod calculated match.

Pevear et al. (1991) and Tsipursky et al.(1992) suggested that kaolinite or similar material may coat illite surfaces, possibly making them behave somewhat like mixed-layer illite/kaolinite. The peak positions for modeled I/K with 5% Kaolinite almost perfectly match many natural illites. This problem has also been addressed by Lanson and Velde (1992) who had similar problems with the calculated illite 001 peak during decomposition analysis. As these authors appealed: we will be impeded from further progress in the XRD petrology of illite until we can fully model the illite 001 peak.

DISCUSSION AND CONCLUSIONS

Factors affecting fit

MatchMod always finds a very good fit (both precise and accurate) to synthetic patterns such as Figure 9, that is, to patterns generated by the same algorithm that is subsequently used to make the fit. This suggests that the GA its doing its job almost perfectly in terms of inverting NEWMOD©. Lack of a good fit must, therefore, be due to the experimental (real) sample containing features not considered by MatchMod, or to improper site settings; a number of these features have been discussed above, a general discussion follows.

Phase not in MatchMod. It is remarkable how many XRD patterns of clays from sedimentary rocks can be approximated by MatchMod's simple four phase system. However, samples containing chlorite/smectite, chlorite/serpentine, vermiculite, sepiolite, palygorskite, opals, and oxides, among others, cannot at present be fully analyzed. MatchMod will fit chlorite/serpentine as simple chlorite, but would likely ignore the others, or worse, fit them as one of the four modeled phases. MatchMod does not presently do I/S with water instead of glycol interlayers, nor the three phase I/S (Illite/1-gly Smectite/2-gly Smectite) reported from shales by Pevear (1992). The latter may result from partial evaporation of glycol during analysis.

Multiple phases of the same broad mineral group. MatchMod can model discrete smectite as I/S with 0% illite layers, but if both discrete smectite and I/S are present, we are stuck. The same applies to samples with two or more I/S phases. Mica and illite in the same sample will be modeled as illite with a large spread of N values, but we cannot differentiate them as Lanson and Champion (1991) and Lanson and Velde (1992) try to do with their peak decomposition method. Using a narrow grain-size range, as was done here, somewhat reduces this problem. Mixtures containing both muscovite and biotite will model as illite with high iron; the same applies to glauconite-containing samples.

Simplistic mineral descriptions in the model. The problem with the illite 001 peak was discussed above; perhaps we need a more sophisticated description of the illite structure. The presence of interlayer Na or NH_4^+ could also alter the illite pattern, but not like the one shown in Figure 13. MatchMod, like NEWMOD©, can be adjusted by user to approximately compensate for interlayer substitution. The Ergun defect broadening option in NEWMOD© (Reynolds, 1989b) is not presently operational in MatchMod; its use will make the peak shape of discrete phases more realistic. Certainly the great variety of structural complexity reported in TEM studies of shales is only very crudely approximated by MatchMod's four simple phases.

Instrument and sample parameters (site settings) incorrect. The slit configuration is especially critical to successful modeling of the pattern over a wide 2θ range; the configuration in MatchMod and in the experimental diffractometer must be the same. Sample length and thickness have been discussed above. Sample displacement errors (Bish and Reynolds, 1989), possibly evident in Figure 13, are not correctable within MatchMod. Mu star is approximated by a typical value in the site settings, but can be measured (Reynolds, 1989c) and set, especially if high iron samples are being run. This is strictly necessary only if absolute intensities are required.

Quantitative analysis

Ideally MatchMod provides an accurate quantitative clay mineral phase analysis including details of phase composition and structure. There are, however, many pitfalls; these are thoroughly discussed by Reynolds (1989c) and in the foregoing. Note that any minerals present other than the four MatchMod phases are ignored. The precision (reproducibility) of MatchMod is generally within 1 to 3 percentage points; the accuracy is unknown except for calculated examples like Figure 9, for which it is 1-3 percentage points as well. There is certainly a question of the uniqueness of any solution, considering the stochastic nature of MatchMod. We can only say that the GA appears to efficiently search the variables, since it always solves the calculated cases properly.

Future directions

Given the apparent success of the simple four phase system, it is tempting to consider versions of MatchMod with more simultaneous phases; for example, mica and illite, or two I/S phases. Incorporation of 3-phase mixed-layering (as in NEWMOD3C©, Reynolds, 1988) is also possible. However, as we add more complexity we add more variables, increasing run time and reducing the likelihood that a match represents a unique solution. We might also consider addressing the probability structure in NEWMOD© directly with the GA, rather than using Reichweite. The long run time (sometimes hours) on a microcomputer is a drawback to MatchMod. We do runs at night on several networked computers operated from a central control using "Timbuktu" (Farallon Computing, Inc.). A much faster workstation version is in progress. MatchMod is ideally suited for parallel processing; as multiprocessor machines become more common, they will be the obvious choices for GA based applications.

ACKNOWLEDGEMENTS

We thank Ray Ferrell, Wuu-Liang Huang, Bob Reynolds, and Jeff Walker for reviews of the manuscript, and our employer for permission to publish it.

REFERENCES CITED

Bish, D. L. and Reynolds, R. C. (1989) Sample Preparation for X-ray Diffraction: in *Modern Powder Diffraction, Reviews in Mineralogy,* **20**, D.L. Bish and J.E. Post, eds., Mineralogical Society of America, Washington, D.C., 73-99.

Goldberg, D. E. (1989) *Genetic Algorithms in Search, Optimization, and Machine Learning*: Addison-Wesley, Reading, MA, 412pp.

Holland, J. H. (1975) *Adaptation in Natural and Artificial Systems:* The University of Michigan Press, Ann Arbor, MI.

Holland, J. H. (1992) Genetic Algorithms: *Scientific American*, **267**, no. 1, 66-72.

Klimentidis, R. E., Huang, W-L., and Pevear, D. R. (1991) Correlation of X.R.D. Crystallinity and T.E.M. Crystal Size of Synthetic Mica: *Program and Abstracts for Clay Minerals Society 28th Annual Meeting, Lunar and Planetary Institute Contribution no. 773*, Houston, TX, 89.

Lanson, B. and Champion, D. (1991) The I/S-to-Illite Reaction in the Late Stage Diagenesis: *American Journal of Science*, **291**, 473-506.

Lanson, B. and Velde, B. (1992) Decomposition of X-ray Diffraction Patterns: a Convenient way to Describe Complex I/S Diagenetic Evolution: *Clays & Clay Minerals* **40**, 629-643.

Moore, D. M. and Reynolds, R. C. (1989) *X-Ray Diffraction and the Identification and Analysis of Clay Minerals:* Oxford University Press, Oxford, 332pp.

Pevear, D. R., Klimentidis, R. E., and Robinson, G. A. (1991) Genetic Significance of Kaolinite Nucleation and Growth on Pre-existing Mica in Sandstones and Shales: *Program and Abstracts for Clay Minerals Society 28th Annual Meeting, Lunar and Planetary Institute Contribution no. 773,* Houston, TX, 125.

Pevear, D. R. (1992) Illite age analysis, a new tool for basin thermal analysis: in *Water-Rock Interaction*, Y. K. Kharaka and A. S. Maest, eds., A. A. Balkema, Rotterdam, 1251-1254.

Post, J. E., and Bish, D. L. (1989) Rietveld Refinement of Crystal Structures Using X-ray Diffraction Data: *Modern Powder Diffraction, Reviews in Mineralogy,* **20**, D.L. Bish and J.E. Post, eds., Mineralogical Society of America, Washington, D.C., 277-308.

Reynolds, R. C. (1980) Interstratified Clay Minerals: in *Crystal Structures of Clay Minerals and their X-ray Identification*, G. W. Brindley and G. Brown, eds., Mineralogical Society, London, 249-303.

Reynolds, R. C. (1983) Calculation of absolute diffraction intensities for mixed-layered clays: *Clays & Clay Minerals* 31, 233-234.

Reynolds, R. C. (1985a) *NEWMOD©, a Computer Program for the Calculation of One-Dimensional Diffraction Patterns of Mixed-Layered Clays.* R.C. Reynolds, 8 Brook Rd, Hanover, NH 03755 U.S.A.

Reynolds, R. C. (1985b) *CLAYS©, a Computer Program for the Calculation of One-Dimensional Diffraction Patterns of Discrete Clays.* R.C. Reynolds, 8 Brook Rd, Hanover, NH 03755

U.S.A.

Reynolds, R. C. (1985c) *MIXER©, a Computer Program for the Addition of X-ray Patterns.* R.C. Reynolds, 8 Brook Rd, Hanover, NH 03755 U.S.A.

Reynolds, R. C. (1986) The Lorentz factor and preferred orientation in oriented clay aggregates: *Clays & Clay Minerals* 34, 359-367.

Reynolds, R. C. (1988) *NEWMOD3C©, a Computer Program for the Calculation of One-Dimensional Diffraction Patterns of 3-Component Mixed-Layered Clays.* R.C. Reynolds, 8 Brook Rd, Hanover, NH 03755 U.S.A.

Reynolds, R. C. (1989a) Principles of Powder Diffraction: in *Modern Powder Diffraction, Reviews in Mineralogy,* 20, D.L. Bish and J.E. Post, eds., Mineralogical Society of America, Washington, D.C., 1-17.

Reynolds, R. C. (1989b) Diffraction by Small and Disordered Crystals: in *Modern Powder Diffraction, Reviews in Mineralogy,* 20, D.L. Bish and J.E. Post, eds., Mineralogical Society of America, Washington, D.C., 145-181.

Reynolds, R. C. (1989c) Principles and Techniques of Quantitative Analysis of Clay Minerals by X-ray Powder Diffraction: in *CMS Workshop Lectures, Vol. 1, Quantitative Mineral Analysis of Clays,* D. R. Pevear and F. A. Mumpton, eds., The Clay Minerals Society, Boulder CO, 3-36.

Snyder, R. L., and Bish, D. L. (1989) Quantitative Analysis: in *Modern Powder Diffraction, Reviews in Mineralogy,* 20, D.L. Bish and J.E. Post, eds., Mineralogical Society of America, Washington, D.C., 101-144.

Tsipursky, S. J., Eberl, D. D., and Buseck, P. R. (1992) Unusual Tops (Bottoms?) of Particles of 1M Illite from the Silverton Caldera (CO): *Agronomy Abstracts, 1992 Annual Meetings,* American Society of Agronomy, Madison, WI, 381-382.

THREE-DIMENSIONAL POWDER X-RAY DIFFRACTION FROM DISORDERED ILLITE: SIMULATION AND INTERPRETATION OF THE DIFFRACTION PATTERNS

R. C. Reynolds Jr.

CONTENTS

Introduction	44
Calculation of Three-Dimensional Powder X-ray Diffraction Patterns	44
Mixed-Layered Illite/Smectite	45
Layer Types	45
Experimental	46
Powder X-Ray Diffraction By Ordered Crystals	46
The Intensity in One Dimension	46
Diffracted Intensity in Three Dimensions	48
Transformation to Orthogonal Reciprocal Space	49
The Orthogonal Reciprocal Lattice	50
X-Ray Powder Diffraction By Disordered Crystals	54
Turbostratic Disorder	55
Disorder Caused by $n.60$ or $n.120$ Degree Rotations of 2:1 Layers	55
Overall Diffraction Equation for Rotationally Disordered Micas--a Summary	59
Treatment of I/S	60
Calculated Patterns	61
Input Model Parameters	61
Interpretation Of Diffraction Patterns	62
General Principles	62
Cv and Tv $1M$ Structures	63
Disordered Cv Structures	64
Disordered Tv Structures	64
Interstratified Cv and Tv Structures	64
Turbostratic Disorder--the Effects of Very Thin Crystallites	64
$n.60°$ Disorder	67
Comparisons Between Calculated And Experimental Diffraction Patterns	68
I/S with Low Expandability	69
I/S with Intermediate Expandability	69
I/S with High Expandability	72
Comments	74
Acknowledgements	76
References Cited	76

THREE-DIMENSIONAL X-RAY POWDER DIFFRACTION FROM DISORDERED ILLITE: SIMULATION AND INTERPRETATION OF THE DIFFRACTION PATTERNS

R. C. Reynolds Jr.

Department of Earth Sciences
Dartmouth College, Hanover New Hampshire 03755

INTRODUCTION

Calculation of Three-Dimensional Powder X-ray Diffraction Patterns

Plançon and Tchoubar (1977a) developed a computer algorithm for calculating three-dimensional powder diffraction patterns of phyllosilicates with various types and degrees of disorder, two of which are treated herein. Turbostratic disorder is caused by randomly distributed random magnitude translations and rotations of the 2:1 silicate layers. This condition breaks the crystal up into thin (along Z) coherent diffracting domains which at the limit of disorder produces an assemblage of crystallites composed of single silicate layers. Many smectites have such a structure and it gives rise to the well-known two-dimensional diffraction bands whose heads are defined by reflections of the type $hk0$ (Brindley, 1980).

Another type of disorder involves rotations of adjacent silicate layers about an axis normal to $d(001)$ and including the center of a K ion in micas. The rotations, which are randomly distributed along Z, may be integral multiples of 60 ($n.60$) or 120° ($n.120$). The structure of a pair of adjacent silicate surface oxygen planes, one of which has been rotated, retains the hexagonal symmetry of a single plane, so this type of disorder is not as drastic as the turbostratic type. Méring (1975) has appropriately termed such structures "semi-ordered", and they are recognized from powder X-ray diffraction patterns by the presence of broad lines for which $k \neq 3n$ and sharp $k = 3n$ peaks. The pure $1M$ mica structure has no rotational disorder, but a structure with a large number of non-coincident layer orientations is termed $1M_d$. Mixed-layering (e.g. illite/smectite) is another type of disorder, but its effects are limited to the basal or $00l$ diffraction pattern.

In a series of definitive papers, stacking disorder in kaolinites was identified and quantified (Plançon and Tchoubar, 1977b; Plançon, 1981; Plançon, et al., 1988). In addition, studies with coworkers investigated the three-dimensional structures of smectites subjected to alkali fixation and repeated wetting and drying (Plançon et al, 1979; Besson et al, 1983; Drits et al., 1984). Drits et al. (1984) calculated diffraction patterns for stacking disordered illite, and Sakharov et al. (1990), using a modified version of the Plançon-Tchoubar algorithm, demonstrated the occurrence of $n.60°$ rotational disorder in glauconite. Drits and Tchoubar (1990) recently published an excellent and comprehensive treatment of powder X-ray diffraction and disorder in layered structures—a book that is the present state-of-the-art summary of the subject.

A more simplified approach is presented here. The present algorithm was constructed from basic optical principles described by James (1965) and the writer's experience with diffraction

Three Dimensional Diffraction from Disordered Illite

from one-dimensionally disordered clay minerals (Reynolds, 1980). Reynolds (1989a, 1989b) discussed the basic principles which are developed further below. The calculus is sufficient to define the algorithm, but the mathematics is minimal because the integrations have no analytical solutions and must be solved numerically. Although no claim is made that this is the most efficient way do the calculations, it is a valid alternative based on the following tests:

1. Peak positions and integrated intensities for calculated three-dimensional diffraction patterns of ordered structures agree with values calculated by

$$I(hkl) = LP \ |F(hkl)|^2,$$

where LP is the random powder Lorentz-polarization factor and F is the structure factor for a single unit cell.

2. Calculated patterns for disordered structures agree with published data (Drits et al., 1984) that used the same disorder parameters and atomic coordinates.

Mixed-Layered Illite/Smectite

Mixed-layered illite/smectites (I/S) from K-bentonites are excellent choices to illustrate the diffraction characteristics of illite structures with different kinds and amounts of disorder because many natural samples are available which are free from peak interferences caused by other minerals that are invariably present in sandstones and shales. In theory, the I/S structure could lead to complicated three-dimensional diffraction models because of variations in mean c with respect to percent expandability and corresponding uncertainties in the three-dimensional atomic structures of the interlayer solvates. However, these complications apparently do not occur in minerals and can be ignored, as experimental work has shown.

Reynolds (1990) reported that air-dried and ethylene glycol-solvated preparations of an I/S mineral gave identical three-dimensional powder diffraction patterns. This result was confirmed on other I/S specimens analyzed in the air-dried, glycol-solvated, and dehydrated (250°C) conditions. Because $d(hkl)$ is independent of mean c for a mixed-layered mineral, no three-dimensional coherence occurs across the expandable interfaces that separate illite layers or thin stacks of illite layers. Each expandable interlayer is a turbostratic defect. Thus, a randomly oriented I/S powder diffracts in three dimensions as a randomly oriented aggregate of exceedingly thin illite crystallites. These thin illite stacks correspond to the illite fundamental particles of Nadeau et al. (1984). Regardless of their genesis, thin illite stacks associated with turbostratic defects and fundamental particles diffract identically in an X-ray diffraction experiment (Reynolds, 1992). The distribution of thicknesses of the illite stacks or fundamental particles is controlled by the percent expendability (% Exp) and the Reichweite (R) which can be determined by X-ray diffraction studies of ethylene glycol-solvated oriented aggregates.

Layer Types

Two types of illite unit cells have been identified in illite and I/S. The best-known is the *trans*-vacant (*tv*) structure with the octahedral cation vacancy in the $M1$ site located on the mirror plane of space group $C2/m$. Zvyagin et al. (1985), however, reported a dioctahedral potassium mica with an occupied $M1$ site, and Tsipursky and Drits (1984) concluded that many dioctahedral smectites contain significant proportions of this noncentric cis-vacant type (cv) thus confirming earlier conclusions of Méring and Oberlin (1967). Drits et al. (1984) published calculated three-dimensional powder X-ray diffraction patterns for cv $1M$ and $1M_d$ illite and demonstrated their differences from the tv type. Reynolds and Thomson (1993) identified cv $1M$ illite from the Potsdam Sandstone of New York, and as shown below, the cv type is well represented in I/S both as a pure end member and interstratified with tv 2:1 layers.

Cartesian atomic coordinates and unit cell parameters for the cv and tv unit cells used in this study are from Drits et al. (1984) and correspond to their designations of C and T layers respectively. Both unit cells have the same lengths $a = 5.199$ Å and $b = 9.005$ Å, but the cv cell has $\beta = 99.13°$, $c = 10.09$ Å, whereas for the tv cell, $\beta = 101.3°$ and $c = 10.164$ Å. Atomic scattering factors were computed following Wright (1973) and corrected for isotropic thermal vibration using $B = 1.5$ for cations and 2.0 for anions.

Calculated diffraction patterns for cv structures incorporate only the real part of the structure factor. The sine series was eliminated thereby defining 2:1 layers with randomly distributed equal proportions of the two possible cv enantiomorphs (a vacancy in either the two cis sites). Although perhaps not physically accurate, this model is mathematically valid because such a disordered layer is statistically centrosymmetric; furthermore, it produced better agreement with experimental diffraction patterns than model structures composed of (1) interstratified layers of each of the two enantiomorphs, and (2) three-dimensional crystals composed of a single enantiomorph.

Cv and tv $3T$ illite (mica) structures have very similar diffraction patterns. They can be distinguished if little disorder is present, but definitive identification may not be possible for I/S because these minerals are usually sufficiently disordered to preclude accurate determinations of d for the $k \neq 3n$ reflections (see Reynolds, 1992). Here they are tentatively interpreted as cv $1M_d$ polytypes.

EXPERIMENTAL

Random powder three-dimensional diffraction patterns for I/S were prepared by side-loading freeze-dried $< 1\mu m$ (equivalent spherical diamter) powders. Sample preparation methods and instrumental operating conditions were given by Reynolds (1992). Powders were heated at 250°C for 1 hr to dehydrate expandable interlayers after studies showed small changes in diffraction patterns for some samples after heating to 350°C, and higher temperatures caused significant structural changes. Ethylene glycol-solvation was not used because, for such preparations, some mixed-layered $00l$ reflections interfere with three-dimensional peaks. Rehydration during analysis was precluded by streaming tank nitrogen through an enclosed sample chamber fitted with a suitable circular Saran Wrap™ window.

POWDER X-RAY DIFFRACTION BY ORDERED CRYSTALS

The Intensity in One Dimension

The one-dimensional Laue interference (ϕ) function (James, 1965) gives the intensity of scattering as a function of h for a crystal containing N unit cells along the X direction, each of which, for example, consists of a single electron

$$\phi(h) = \frac{\sin(\pi N h)^2}{N \sin(\pi h)^2} . \qquad (1)$$

Infinitely thick crystals diffract only at integral values of h, but small and/or disordered crystals have broadened reflections distributed over a range in h, requiring that h be treated as a continuous variable.

Figure 1 shows the interference function for the 100 reflection computed for $N = 10$. The maximum repeats at integral values of h, and there are $N - 2$ equally spaced ripples between the maximal heights of N. The integrated area between $h_0 \pm 2/N$ is 0.96, and the breadth (in 2θ space) at one-half maximal height is inversely proportional to N according to the Scherrer equation

$$\beta_{1/2} = \frac{K \lambda}{t \cos \theta} ,$$

where $\beta_{1/2}$ is the peak width at half-height in 2θ (radians), K is a constant close to unity, λ is the wavelength, and t is the thickness in Å ($N \cdot d(100)$) of the crystallite in a direction normal to the diffraction planes 100.

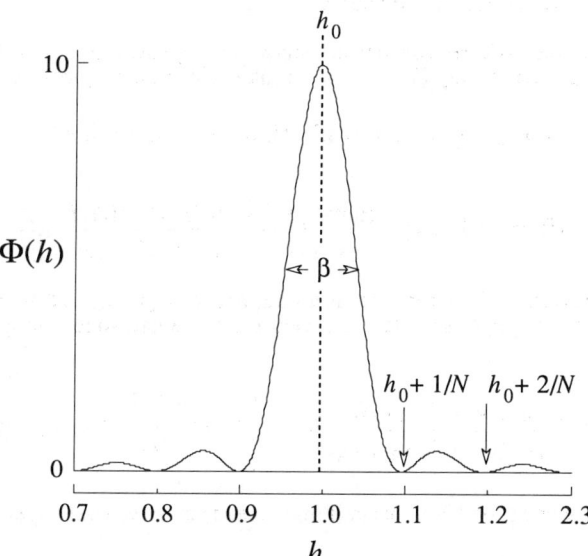

Figure 1. The one-dimensional interference function. Small maxima are ripples which cancel out with a distribution of crystallite sizes.

Real crystals are more complicated, and for a unit cell with a center of symmetry, the one-dimensional scattering amplitude is given by the structure factor

$$|F(h)| = \sum_j n(j) f(j) \cos(2\pi\, h\, x(j)), \qquad (2)$$

where $n(j)$ is the site occupancy at $x(j)$, and $f(j)$ is the temperature-corrected atomic scattering factor. The summation is taken over one-half of the unit cell from the origin $j = 0$ (the center of symmetry). The fractional atomic coordinate $x(j)$ is equal to the separation of a given atomic position (in Å) from the origin of the calculation divided by the unit cell dimension a. The complete diffraction equation is

$$I(h) = LP\, |F(h)|^2\, \phi(h)\ . \qquad (3)$$

LP is the Lorentz-polarization factor, which for a point-by-point calculation of the diffraction pattern for a random powder is

$$LP = \frac{1 + \cos^2(2\theta)}{\sin^2(\theta)}\ . \qquad (4)$$

Diffracted Intensity in Three Dimensions

Extension to three-dimensional diffraction requires an expanded structure factor equation (James, 1965) and the product of Eq. (1) with two similar sine-squared quotients.

$$|F(hkl)| = \sum_j n(j) f(j) \cos(2\pi\, (h\, x(j) + k\, y(j) + l\, z(j)))\ , \qquad (5)$$

$$I(hkl) = LP\, |F(hkl)|^2\, \frac{\sin(\pi N_1 h)^2}{N_1 \sin(\pi h)^2}\, \frac{\sin(\pi N_2 k)^2}{N_2 \sin(\pi k)^2}\, \frac{\sin(\pi N_3 l)^2}{N_3 \sin(\pi l)^2}\ . \qquad (6)$$

The interference function for l is introduced as a Fourier series (James 1965) to accommodate disordered arrangements of phyllosilicate unit layers and to transform to orthogonal reciprocal space,

$$I(hkl) = LP\, |F(hkl)|^2\, \frac{\sin(\pi N_1 h)^2}{N_1 \sin(\pi h)^2}\, \frac{\sin(\pi N_2 k)^2}{N_2 \sin(\pi k)^2}\, \frac{1}{N_3}\left[N_3 + 2 \sum_{n=1}^{n=N_3-1} (N_3 - n) \cos(2\pi\, n\, l) \right]. \qquad (7)$$

The sum is over n, the number of 10-Å spacings that separate any two mica layer centers.

The intensities at a sequence of 2θ values produce a random powder diffraction pattern, and to calculate the intensity at one angular increment, a summation of Eq. (7) is required over those values of h, k, and l for which the diffraction angle (or d) is identical. A solution requires

Three Dimensional Diffraction from Disordered Illite

Figure 7a depicts one idealized hypothetical array out of many which define a disordered structure. The number of interlayers (n) is 5, and with the three "components" 0, 120, and 240° rotated layers, the total number of arrays required is $3^5 = 243$. The hexagons represent the surface oxygen planes of the top tetrahedral sheets for each of the 2:1 layers in the stacking array. The intersection of the surface oxygen plane with the unit cell mirror is shown by the line that bisects each hexagon. Each of the upper oxygen planes is displaced from the equivalent plane of its neighbor by the unit cell stagger along X, which is $a_0 = |\,c \cos \beta\,| = a/3$ for micas with the ideal monoclinic angle.

Figure 7b shows corresponding stacking vectors and the shift of the final layer with respect to the first. For all arrays, the first 2:1 layer is oriented coincident with the *1M* setting, and the origin of the calculation is located where the plane of the octahedral cations intersects a line connecting the K atoms at the top and bottom of the unit cell. That position is indicated by a circle at the center of each stacking vector. The magnitudes of the translation vectors along X are a_0 for zero and 180 degree rotations, and $|a_0 \cos 120|$ for rotations of 60, 120, 240 and 300°. The magnitudes of the translations along Y are $|a_0 \sin 120|$ for rotations of 60, 120, 240, and 300°, and zero for zero and 180° rotations. The overall displacements along X and Y between the two end layers (Figure 7) are computed by summing the shifts for layers with zero, 120 and 240° azimuthal orientations. The stacking array shown in Figure 7 has the following sequence of rotations (starting with the second layer): 0 -- 120 -- 240 -- 120 -- 0, so

$$\Delta X = -a_0 - a_0 + a_0 \cdot 0.5 - a_0 + a_0 \cdot 0.5 + a_0 \cdot 0.5 - (a_0 \cdot 0.5 / 2 - a_0 / 2) ,$$

and

$$\Delta Y = 0 + 0 + 0.866 \cdot a_0 + 0 + 0.866 \cdot a_0 + 0.866 \cdot a_0 - (0.866 \cdot a_0 / 2) .$$

Quantities in parentheses at the end of each equation correct for end effects that result from locating the origins of the layers at their centers.

The phase angle for the separation of the end layers is $2\pi\,(h\,\Delta X\,/\,a + k\,\Delta Y\,/\,b)$, and this is substituted into the argument of the cosine term of Eq. (14) to give

$$\cos\,(2\pi\,(h\,\Delta X\,/\,a + k\,\Delta Y\,/\,b + nl)) . \qquad (16)$$

The probability of occurrence of this array is

$$\sigma_{120} = P_0^2 \cdot P_{120}^2 \cdot P_{240}^1 .$$

The exponents refer to the number of zero, 120 and 240° interfacial rotations in the array and the subscript of σ means that the successive rotations have resulted in a top layer with an azimuthal orientation of 120°. The first layer is ignored in the probability calculation because all arrays start with a non-rotated layer, and to include a value for it would simply multiply the probabilities of all arrays by a constant.

A layer not coincident with the *1M* setting is a separate component with a unique structure factor. Consider the structure factor product between layers with zero and 120° rotations (Figure 8). The structure factor for the 120° component is computed by means of Eq. (5) using transformed values of h and k, in this case, the transformation of the $02l$ to the $\overline{11}l$. This procedure requires that F' is based on orthogonal atomic coordinates, that it is restricted to integral values of h and k, and that the unit cell is strictly orthohexagonal, that is, $b = \sqrt{3}\ a$.

One term of the summation in Eq. (14) is given by Eq. (17) where $F'_{120}(h_0\ k_0 l)$ refers to the structure factor for a unit layer that is rotated 120° from the $1M$ setting. Equation (17)

$$(N_3 - 5)\ \sigma_{120}\ F'_0(h_0\ k_0 l)\ F'_{120}(h_0 k_0 l)\ \cos(2\pi\ (h_0\ \Delta X\ /\ a + k_0\ \Delta Y\ /\ b + n\ l\ )) \quad (17)$$

presents formidable difficulties for any practical solution. A very large number of arrays similar to the one illustrated by Figure 7 are required. Their application, one by one, in the summation within the numerical integration causes prohibitive run times for any computer. In addition, this procedure must be repeated for each 2θ increment of the diffraction pattern. The difficulty is eliminated by making use of the identity

$$\cos(x + y) = \cos(x)\cos(y) - \sin(x)\sin(y).$$

Let

$$x = 2\pi\ (h_0\ \Delta X\ /\ a + k_0\ \Delta Y\ /\ b)$$

and

$$y = 2\pi\ n\ l.$$

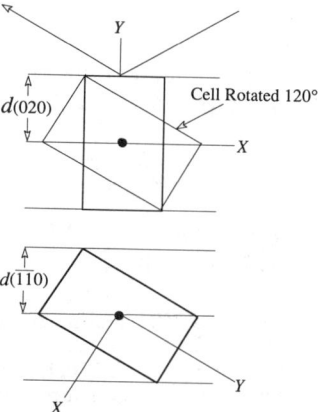

Figure 8. Transformation of the structure factor for a rotated layer.

Equation (17) is rewritten

$$(N_3 - 5)\, \sigma_{120}\, F'_0(h_0k_0l)\, F'_{120}(h_0k_0l) \cdot$$

$$(\cos(2\pi\, (h_0\, \Delta X / a + k_0\, \Delta Y / b))\, \cos(2\pi\, n\, l) -$$

$$\sin(2\pi\, (h_0\, \Delta X / a + k_0\, \Delta Y / b))\, \sin(2\pi\, n\, l)). \tag{18}$$

The sine and cosine terms for each h_0k_0 rod are independent of d (or 2θ) and l and are evaluated outside of the summation and the integral over g_l, allowing a very large reduction in computation time. Each trigonometric term is multiplied by the corresponding σ for that array, and the probabilities and phase shifts are combined in matrices that are calculated separately from the computation of the diffraction pattern. These matrices, $C(n,h_0k_0,\omega)$ and $S(n,h_0k_0,\omega)$, contain solutions for the sums of the products of the array probabilities times the cosines and sines respectively of the phase shifts between the initial and final layers in the arrays. The summation of Eq. (19) requires values for arrays that contain $1, 2, 3\cdots n = N_3 - 1$ interlayers. The angle ω is the azimuthal orientation of the terminating layer, and this quantity is required for the structure factor, F', that applies to a layer rotated ω degrees from the azimuthal reference direction (Eq. (19)). Finally, the indices h_0k_0 identify the rods intersected by the Ewald sphere. The details of the algorithm for these computations are beyond the scope of this paper, but the calculation must utilize recursion (see Bethke and Reynolds, 1986) because the number of arrays is too large for the generation and incorporation of each into the matrices. For example, structures composed of interstratified cv and tv layers with rotations of 0, 60, 120, 180, 240, and 300° are examples of twelve-component interstratification, and for a realistic thickness along Z of 10 unit cells, some elements of the matrices are the sums of 12^{10} or about 6.2×10^{10} individual arrays, this number describing all possible combinations and permutations of twelve entities taken ten at a time. The number increases twelve-fold with each additional layer, but the recursive algorithm employed here causes only a two-fold increase in the computation time.

Overall Diffraction Equation for Rotationally Disordered Micas--a Summary

The following equation includes the substitutions and simplifications described above that separate it from Eq. (9).

$$I(1/d) = \frac{LP}{N_3} \int_{gl} R_1 \left[\overline{N}_3\, |F_0(h_0k_0l)|^2 + 2 \sum_{n=1}^{n=N_3-1} M(n) \cdot \right.$$

$$\left. \left[\sum_\omega F_0(h_0k_0l)\, F_\omega(h_0k_0l) \left(\cos(2\pi nl)\, C(n,h_0k_0,\omega) - \sin(2\pi nl)\, S(n,h_0k_0,\omega) \right) \right] \right].$$

$$\int_{g_{hk1}}^{g_{hk2}} \frac{\sin(\pi N_1 h)^2}{N_1 \sin(\pi h)^2} \frac{\sin(\pi N_2 k)^2}{N_2 \sin(\pi k)^2}\, d_l\, dg_{hk}. \tag{19}$$

To preserve compactness, however, Eqs. 10, 11, 12 and 13 for R_1, h, k, and l have not been incorporated and these must be considered for the final result. Equation 19 gives a solution for one of the rods h_0k_0; it must be summed for up to eleven rods to produce the powder diffraction intensity at one value of $1/d$ or 2θ.

Treatment of I/S

The different coherence conditions for basal and three-dimensional diffraction for I/S require different crystallite thickness distributions for each. A computer program was written similar to the one described by Altaner and Bethke (1988) that generates all combinations and permutations of illite and expandable interfaces between 2:1 layers in MacEwan crystallites (Altaner and Bethke, 1988) for which $N = 1, 2, 3\cdots N_3$ silicate layers. The frequency of occurrence for each N is calculated by Markovian statistics (Reynolds, 1980; Bethke and Reynolds, 1986) and weighted by the Ergun (1970) defect broadening quantity $\exp(-(N-1)/\delta)$ where δ is the mean defect-free distance. This distance, δ, was set equal to 6 because this value commonly occurs in successful models of illite basal reflections. For each MacEwan crystallite, weighted by its probability of occurrence, the numbers of 10, 20, 30, 40-layer etc. illite particles (or illite stacks or fundamental particles) are counted and the sums pooled with the results for the next crystallite and then for the crystallites for the next N. This procedure produces a particle thickness distribution similar to the fundamental particle histograms of Nadeau et al. (1984).

Crystallite end effects produce single layers ($N = 1$) for R > 0 structures (see also Altaner and Bethke, 1988). These are probably unrealistic and represent an inadequacy of the Markov approach for R > 0 structures. The single layers were eliminated and the crystallite size distributions were renormalized to a sum of unity. Figure 15 to 24, shown below, compare experimental and calculated diffraction patterns for I/S minerals with % Exp and R. The validity of this approach is confirmed by the successful diffraction models for all of the minerals studied here. Table 1 shows calculated three-dimensional crystallite thickness distributions for a range of % Exp and R.

Table 1. Crystallite thickness distribution with respect to Reichweite and percent expandability.

	R3	R1	R1	R1	R0	R0	R0	R0	R0	(Reichweite)
	10	20	30	40	50	60	70	80	90	(% Exp)
N^1			Relative Abundances							
1	0	0	0	0	0.57	0.66	0.75	0.83	0.92	
2	0.07	0.34	0.51	0.72	0.25	0.23	0.19	0.14	0.08	
3	0.19	0.25	0.27	0.21	0.11	0.08	0.05	0.02		
4	0.27	0.19	0.13	0.05	0.05	0.03	0.01			
5	0.19	0.11	0.06	0.01	0.02					
6	0.12	0.06	0.02							
7	0.08	0.04								
8	0.05									
9	0.03									

N^1 is the number of 2:1 layers per crystallite.

Three Dimensional Diffraction from Disordered Illite

The 00l series for I/S is produced by coherent scattering over the entire MacEwan crystallite. The MacEwan crystallite is, therefore, thicker than N_3 for a typical three-dimensional crystallite. The basal series was calculated separately by a subprogram like NEWMOD© (Reynolds, 1985) and added to the three-dimensional calculated pattern, after suitable normalization for absolute intensity, This basal series contains no information on polytypes or disorder--it is included for "cosmetic" purposes and to correctly simulate interferences between three- and one dimensional diffraction peaks.

CALCULATED PATTERNS

A computer program (WILDFIRE) written in True Basic© calculates diffraction patterns by means of Eq. (19). The Macintosh IIfx executes a typical diffraction pattern from 16 to 44 °2θ at increments of 0.1° in about ten minutes, though the most demanding cases involving $N_3 = 10$ and interstratified cv/tv structures with both $n.60$ and $n.120$ rotations can take as long as 30 minutes.

Input Model Parameters

Table 2 lists the variables that define a model structure. Calculated patterns are insensitive to some, a few can be fixed by independent experiments, and the unconstrained variables can be estimated well enough for a first approximation by inspection of an experimental diffraction pattern. The refined pattern requires trial-and-error adjustments of the latter.

N_1 and N_2 are, respectively, the numbers of unit cells in the X and Y directions. The default values of 60 and 30 define a crystal that is approximately a square 300 Å on a side. The models are not sensitive to these variables and all but one of the calculated patterns in this paper used the default values. They would be changed if, for example, a model pattern was required for lath-shaped crystals.

Table 2. Input parameters for model diffraction patterns.

Parameter	Description	Default
N_1	No. of Unit Cells Along X	60
N_2	No. of Unit Cells Along Y	30
Pf	Preferred Orientation	1.0
Fe	No. of Fe atoms/Si_4O_{10}	0.14
K	No. of K atoms/Si_4O_{10}	0.7
δl	Defect-free distance along Z for 00l	6
Nl	Maximum N along Z for 00l	20
N_3	No. of Unit Cells Along Z	Variable[1]
P_0	Fraction of Zero Rotations	Variable[2]
P_{60}	Fraction of 60, 180, 300 Rotations	Variable[2]
P_{cv}	Fraction of cis-vacant layers	Variable[2]

[1] A single value or a distribution of values.
[2] Unconstrained variable.

Pf corrects intensities for preferred orientation according to Dollase (1986). Unity gives random orientation, and lesser values simulate the effects of preferred orientation for the reflection mode of powder diffractometry. *Pf* can be estimated by optimizing the agreement between calculated and experimental patterns for the intensities of the 020; 110 reflections with respect to the 002 reflection. *Pf* lies between 0.9 and 1 for all of the calculated patterns shown herein save one, attesting to the near-random orientation achieved by the experimental methods employed.

Fe and K refer to the number of Fe and K atoms per formula unit. The default value for Fe applies to all calculated patterns. K was decreased with expandability between the limits of 0.7 for illite to 0.3 for smectite for the results shown by Figure 13.

The value of N_3 refers to the number of unit cells along Z and provision is made for a distribution to simulate an assemblage of crystallites. For a non-mixed-layered illite, suitable values can be obtained from the analysis of the shape and breadths of the 00*l* reflections by means of the program NEWMOD©.

P_0 is the probability of a layer rotated zero degrees with respect to the preceding layer. The *1M* polytype is described by $P_0 = 1$. For structures with $n.120°$ disorder, $P_{120} = P_{240}$, and $P_0 = 1/3$ produces equal proportions of interlayers involving 0, 120, and 240° rotations of adjacent layers. Changing P_0 sequentially from 1/3 to 1 describes the transition from $1M_d$ to *1M*. Only random patterns of layer rotations are considered, that is, no nearest-neighbor or next-to-nearest neighbor ordering conditions are treated to describe discrete polytypes (except for *1M*) such as $2M_1$ and *3T*.

Rotations of 60, 180, and 300° are possible, and their abundance is controlled by the variable P_{60}. $P_{60} = 0$ indicates that the disorder is of the type $n.120°$, and setting $P_{60} = 1$ eliminates the rotations 120 and 240° giving random stacking sequences of 60, 180 and 300° rotations. Setting $P_{60} = 0.6$ produces equal proportions of 60, 120, 180, 240, and 300° rotations.

The variable P_{cv} in Table 2 denotes the proportion of *cv* layers interstratified with *tv* layers in a model structure, thus, $P_{cv} + P_{tv}$ equals unity.

All diffraction angles on the figures and in the text apply to CuKα radiation, and the calculated and experimental diffraction patterns described and interpreted herein are I/S in the dehydrated condition with $d(001)$ for smectite equal to 9.7 Å.

INTERPRETATION OF DIFFRACTION PATTERNS

General Principles

Interpretation of random powder patterns should proceed in a systematic fashion. An analysis of mica patterns profitably begins by an analysis of the 02*l*; 11*l* reflections between 19 and 34 °2θ. Their positions and intensities are diagnostic for the various mica polytypes and for the *cv* and *tv* structures. The diagnostic reflections are of the type $k \neq 3n$ and they are broadened by small crystallite size and/or rotational disorder. Evaluating these two alternatives requires consideration of the $k = 3n$ 20*l*; 13*l* peaks between 34 and 39 °2θ which are broadened by small crystallite size but are relatively unaffected by $n.120°$ rotational disorder. The $20l$; $13l$ signature is modified by turbostratic defects and by $n.60°$ rotations as Sakharov et al. (1990) have shown for glauconite.

Cv *and* Tv 1M *Structures*

Figure 9 compares calculated diffraction patterns for *cv* and *tv* partially disordered *1M* or $1M_d$ types ($P_0 = 0.8$). The two are readily distinguished by the position and intensity of the $02l$; $11l$ reflections, although the peaks are broadened by disorder. The two $20l$; $13l$ composite peaks between 35 and 39 °2θ are broader for the *tv 1M* structure because the monoclinic angle ($\beta = 101.3°$) is sufficiently different from the ideal orthohexagonal value that it causes partial resolution of each into its members (Bailey, 1991). For both structures, the $20l$; $13l$ signature is well developed indicating relatively coarse crystallite thickness, that is, few turbostratic interlayers. Note that the 022 reflection is strong for the *cv* structure causing an apparent high peak intensity ratio for the 003/002.

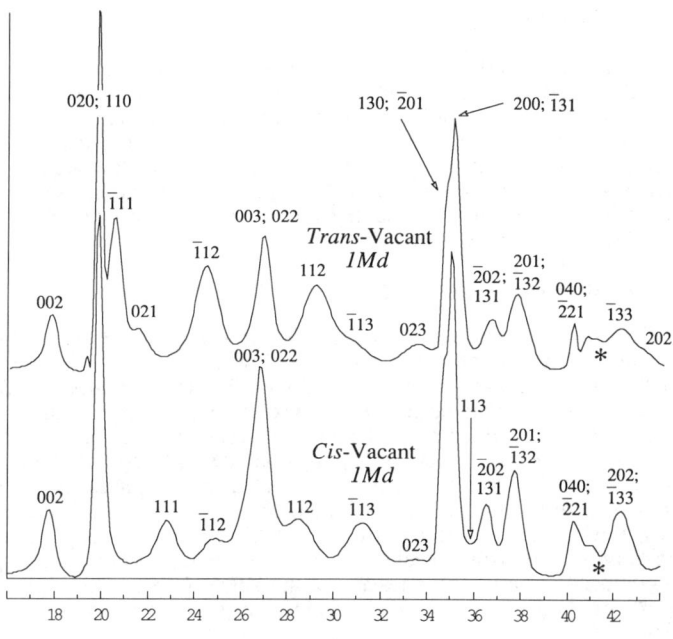

Figure 9. Calculated diffraction patterns for *cv* and *tv* $1M_d$ structures. $N_1, N_2 = 60, 30$; $Pf = 1$; K, Fe = 0.7, 0.14; $\delta l, Nl = 6, 20$; $q(1) -- q(7)$ (Eq. (15)) = 0, 0.068, 0.198, 0.298, 0.205, 0.139, 0.093; $P_{60} = 0$; $P_0 = 0.8$, $P_{cv} = 1$ and $P_{tv} = 1$. Stars refer to the angular region that contains 6 poorly resolved peaks that have not been indexed.

Disordered Cv Structures

Figure 10 demonstrates that $n.120$ rotational disorder causes the demodulation of the $11l$ reflections into a broad background maximum centered about about 26 °2θ, with the $20l$; $13l$ peaks relatively unaffected. These features allow discrimination between rotational disorder and small crystallite size which causes broadening of all reflections. Disorder does not shift peak position, so if other criteria consistent with large crystallite size are evident, the positions of the $11l$ reflections can be used to estimate β and the unit cell dimensions even for I/S.

Disordered Tv Structures

Rotational disorder of the $n.120°$ type has nearly the same effects on the *tv* structure diffraction pattern (Figure 11) as it does for the *cv* variety, except that changes in the $20l$; $13l$ signature arise from the large monoclinic angle of the *tv* unit cell. Note that the $02l$; $11l$ reflections rapidly take on the shape of a turbostratic two-dimensional band with increased disorder, but the $20l$; $13l$ peaks do not, highlighting the difference between manifestations of turbostratic and rotational disorder.

Interstratified Cv and Tv Structures

Equation 16 must be expanded to six components to treat interstratified *cv - tv* structures with $n.60$ or $n.120°$ rotational disorder, or to 12 components for the complicated case of two types of layers and two types of disorder. Calculated diffraction patterns for $n.120°$ disordered interstratified *cv / tv* $1M_d$ ($P_0 = 0.7$) examples are shown in Figure 12 where P_{cv} is the fraction of *cv* layers. The relative intensities of the $11l$ ($l > 0$) reflections allow a good estimate of P_{cv} and the positions show conventional mixed-layer migration with composition. The different peak positions for the two end members arise from the different monoclinic angles for each-- 99.13° for the *cv* and 101.3° for the *tv* unit cells.

Turbostratic Disorder--the Effects of Very Thin Crystallites

Calculated diffraction patterns for I/S with different % Exp and R are given by Figure 13. All of the calculated I/S diffraction patterns are based on equal $n.120°$ disorder, namely, $P_0 = 0.8$, but the K content is decreased from 0.7 to 0.3 with % Exp. The major differences among the patterns are caused by different particle thickness distributions that are, in turn, controlled by % Exp and R (see Table 1). In fact, the fraction of turbostratic interlayers and % Exp are identical. The limit of % Exp = 1 defines a pure smectite with the only "three-dimensional" character being the single 2:1 layer, that is, $N_3 = 1$. Such a mineral produces a diffraction pattern with the two-dimensional bands that signify a completely turbostratic structure (bottom trace, Figure 13). As % Exp increases, thicker "illite" crystallites are more abundant, and modulation of the 20; 13 band into three distinct peaks is evident. Reynolds (1992), utilizing experimental data, showed a comparison between a measure of the $20l$; $13l$ diffraction signature and % Exp for a series of I/S minerals, and that comparison conforms closely to the calculated series displayed in Figure 13.

The effects of crystallite thickness on the $02l$; $11l$ peaks is similar to the effects of rotational disorder in thick crystals--with one important difference. Decreased crystallite thickness not only broadens these reflections, but also shifts them from their "correct" positions. Thus, unlike the

Three Dimensional Diffraction from Disordered Illite

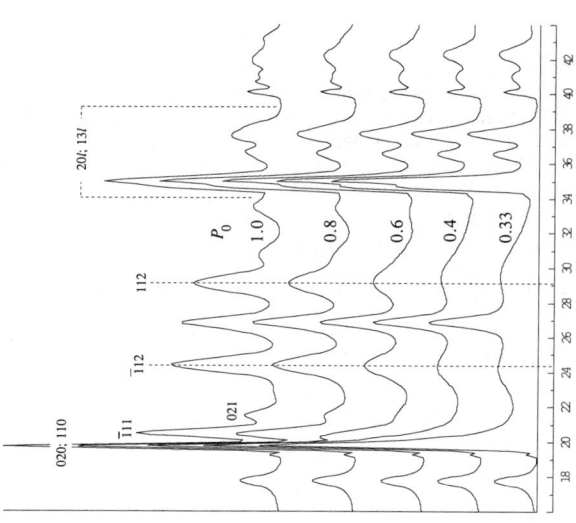

Figure 11. Calculated diffraction patterns for disordered tv $1M$ structures. $P_{60} = 0$; $P_{cv} = 0$; other parameters as in Figure 9.

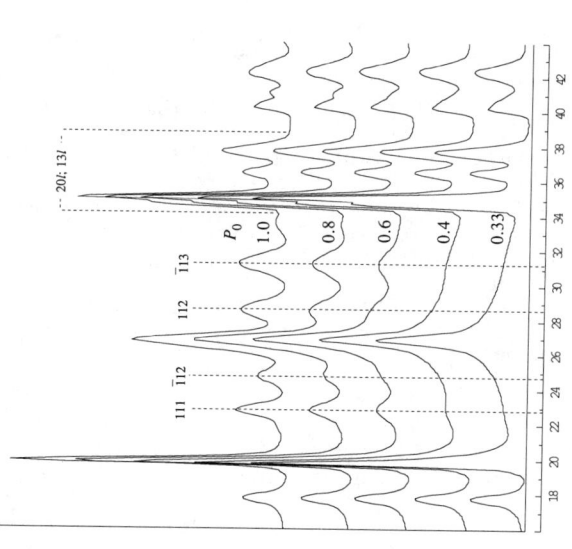

Figure 10. Calculated diffraction patterns for disordered cv $1M$ structures. $P_{60} = 0$; $P_{cv} = 1$; other parameters as in Figure 9.

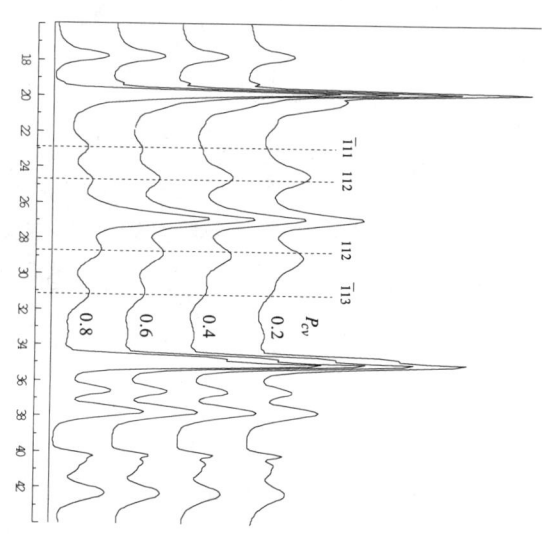

Figure 12. Calculated diffraction patterns for interstratified cv and tv $1Md$ structures. $P_{60} = 0$; $P_{cv} = 0$; $P_0 = 0.7$; other parameters as in Figure 9.

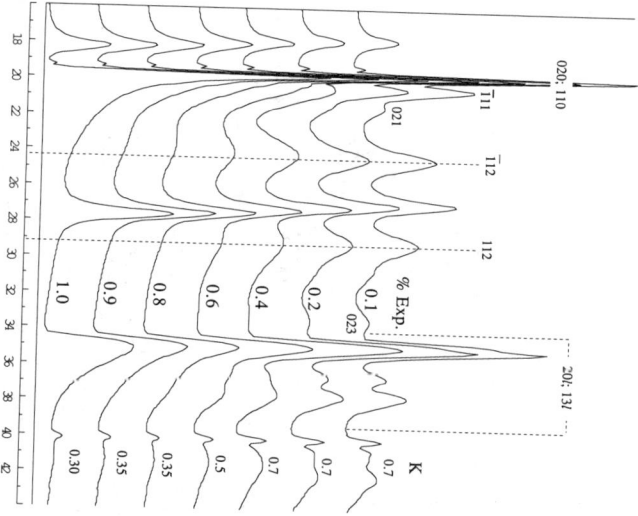

Figure 13. Calculated diffraction patterns for tv $1Md$ illites with different expandabilities. $R = 1$ for % Exp <0.5 and $R = 0$ for values > 0.5. $N_1, N_2 = 60, 30$. $Pf = 1$; Fe = 0.14; $\delta l, Nl = 6, 20$; $q(N)$ (Eq.(15)) from Table 1. $P_{cv} = 0$; $P_0 = 0.8$. Potassium content is labelled on the figure.

case of rotational disorder, they can no longer be used to provide accurate measures of the unit cell dimensions and shape (see also Drits and Tchoubar, 1990). Cv minerals show similar effects of crystallite thickness. Fortunately, rotational disorder affects the 20l; 13l signature much less than crystallite thicknesses for most illite and I/S, so the two types of disorder can be separately quantified by a study of both 02l; 11l and 20l; 13l classes of reflections.

n.60° Disorder

Sakharov et al. (1990) first demonstrated the effects of $n.60°$ disorder on mica diffraction patterns. These rotations have little effect on the 02l; 11l peaks except to sharpen them somewhat, but the transformation of the 20l; 13l reflections is so dramatic that an estimation of their abundance (P_{60}) is easily obtained from a study of the diffraction pattern between 34 and 39 °2θ.

Figure 14 demonstrates the effects of increased $n.60°$ rotations on the 20l; 13l reflections. $P_0 = 0.7$ so 30% of the interlayers separate non-coincident adjacent layers. P_{60} is the fraction of this 30% that consists of equal proportions of 60, 180, or 300° rotations. The calculated diffraction traces show that increased proportions of $n.60°$ disorder increase the symmetry (Figure 14) of the composite reflection near 35 °2θ and cause the migration of the $\bar{2}$02; 131 and 201; $\bar{1}$32 peaks towards each other, therefore, in the limit $P_{60} = 1$ (for this value of P_0), a single broad maximum results that is centered at about 37.2 °2θ.

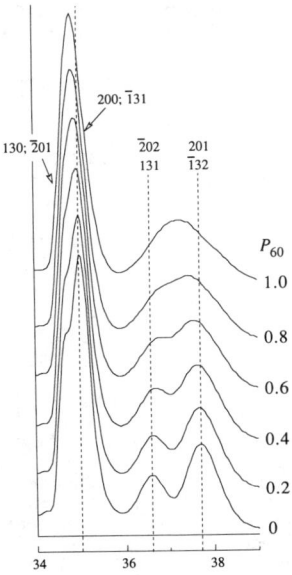

Figure 14. Calculated diffraction patterns demonstrating the effects of $n.60°$ rotational disorder on the 20l; 13l reflections. $P_{cv} = 0$; $P_0 = 0.7$; other parameters as in Figure 9.

The shape of the 20l; 13l diffraction profile allows an estimation of the fraction of $n.60°$ disorder provided that allowance is made for crystallite thickness effects and for total disorder defined by P_0. There are, however, other factors that can complicate such an analysis. Increased tetrahedral tilt increases the intensity of the $\overline{2}02$; 131 with respect to the 201; $\overline{1}32$ reflections, and a higher Fe content has the opposite effect. $n.60°$ disorder still causes migration of the two peaks to form one, but the asymmetry shown at the middle of the range of Figure 14 may be reduced. A unique case occurs for I/S compositions near 50 % Exp, $R = 1$. Almost all of the three-dimensional crystallites are 20 Å thick, and for $P_0 = 1$ they produce a single $\overline{2}02$; 131, 201; $\overline{1}32$ reflection very similar to the one observed for $n.60°$ disorder.

COMPARISONS BETWEEN CALCULATED AND EXPERIMENTAL DIFFRACTION PATTERNS

Figures 15 to 24 provide comparisons between calculated and experimental diffraction patterns selected to illustrate a range in the structural possibilities of I/S. Sample information is given by Table 3. The comparisons are shown (1) to illustrate the principles of interpretation described previously, (2) to demonstrate the power of calculated patterns to realistically model experimental diffraction data from I/S.

For each example, the particle thickness distribution was calculated from % Exp (as was the K content) and R determined by analysis of the ethylene glycol-solvated basal diffraction pattern from an oriented specimen. Values for the mean defect-free or coherence distance (δl) along Z and the maximum distance (Nl) were estimated by modeling 00l peak shapes with NEWMOD©. The preferred orientation value (Pf) was 0.9 or greater for all except the Deicke K-bentonite (KB). Other input parameters were fixed for the series except for the unconstrained variables (Table 2) P_0, P_{cv}, and P_{60} which were adjusted by trial and error to produce the model diffraction patterns. The Mancos A KB from Cerrillos is exceptional. A good deal of "tinkering" was necessary to produce an acceptable model diffraction pattern, so this example inspires less confidence in a unique solution than do the others.

Table 3. Samples studied.

Sample[1]	% Exp; R	Age	Location
Oslo KB	<5; -	Silurian	Oslo, Norway
Glencoe Shale	5; 3	Ordovician	Saint Louis, Missouri
Onandaga KB	10; 3	Devonian	Syracuse, New York
Deicke KB	10; 3	Ordovician	Martinsburg, West Virginia.
Shiphead KB	19; 1	Devonian	Gaspé, Quebec
Mancos A KB	30; 1	Cretaceous	Cerrillos New Mexico
Marias River KB	31; 1	Cretaceous	Sun River Canyon, Montana
Utica KB	40; 1	Ordovician	Quebec City, Quebec
Mancos B KB	54; 0	Cretaceous	Penrose, Colorado
Mancos C KB	75; 0	Cretaceous	Westwater, Utah

[1]KB designates K-bentonite.

I/S with Low Expandability

Figures 15 to 18 show model and experimental I/S diffraction patterns for four samples that are 10% or less expandable. The Oslo KB (Figure 15) has the *tv 1M* structure as indicated by the strong $\bar{1}12$ and 112 reflections and the absence of the diagnostic *cv* 111 peak at about 23° 2θ. Disorder is moderate ($P_0 = 0.55$) because the $\bar{1}12$ and 112 reflections are well-defined but broad, and 60, 180 and 300° rotations ($P_{60} = 0.5$) are frequent producing almost a single peak for the 20*l*; 13*l* reflections between 36 and 39° 2θ.

The Glencoe Shale (Figure 16) provides a good example of a highly disordered structure ($P_0 = 0.33$) with equal proportions of 60, 120, 180, 240, and 300° rotations resulting is a single broad reflection in the 20*l*; 13*l* region. The $\bar{1}12$ and 112 reflections are broadened to the extent that they almost merge with the background maximum at about 27 °2θ which is characteristic of disordered mica structures. Like the Oslo KB, the layer type is primarily *tv*, although a small amount ($P_{cv} = 0.35$) of *cv* layers was necessary to successfully simulate the background between 22 and 26 °2θ. With current nomenclature, the Oslo KB would be labelled *1M* and the Glencoe Shale, $1M_d$.

The diffraction patterns of the Deicke KB (Figure 17) and the Onondaga KB (Figure 18) differ markedly from the previous two because the structures are dominated by *cv* layers and *n*.120° rotations. The ordering parameters (P_0) are similar to the Oslo KB, but the 02*l*; 11*l* reflections produce four peaks of nearly equal intensity (excluding the 022 which is superimposed on the 003 reflection) which is characteristic for the *cv* structure. The Deicke KB, compared to the Onondaga KB, has a higher proportion of *tv* layers ($P_{cv} = 0.6$ vs. 0.8) as indicated by the weaker 111 and 113 compared to the $\bar{1}12$ and 112 reflections. Rotational disorder is mostly of the *n*.120° type, resulting in two well-defined peaks between 36 and 39 °2θ; comparison should be made with the same diffraction angles of Figures 15 and 16 to appreciate the difference that the type of rotational disorder produces in the 20*l*; 13*l* diffraction signature.

The experimental and calculated diffraction patterns are well correlated in every respect save one. Between about 40.2 and 41.5 °2θ both *cv* and *tv* structures produce five weak reflections which merge into a modulated background for any significant amount of disorder. The calculated patterns reproduce this effect poorly, and no explanation is available at present.

I/S with Intermediate Expandability

The Shiphead KB (Figure 19) is similar to the Oslo KB except that interstratified *cv* layers, not disorder, have produced the weaker $\bar{1}12$ and 112 and the suggestion of a weak 111 reflection at 23°. The well-modulated 20*l*; 13*l* maxima indicate a significant proportion of *n*.120° rotations.

Figures 20 to 22 depict diffraction patterns for I/S with 30 - 40% expandability. Illitic stacks are thin, for these compositions compared to the examples discussed previously, producing an increased number of turbostratic defects at interfaces. The effect on the pattern is to increase the intensity of the saddle at 36 °2θ. A measure of this saddle height was called the turbostratic index (TSI) by Reynolds (1992), who discussed it in more detail. Figures 20 to 22 demonstrate an increased TSI with respect to the previous examples (see Figure 13).

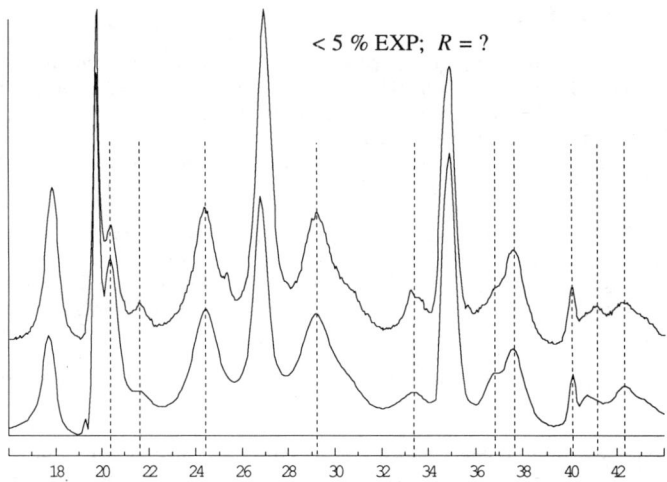

Figure 15. Comparison between calculated (bottom) and experimental diffraction patterns for the Oslo KB. Non-default parameters are $P_0 = 0.55$, $P_{cv} = 0$ and $P_{60} = 0.50$.

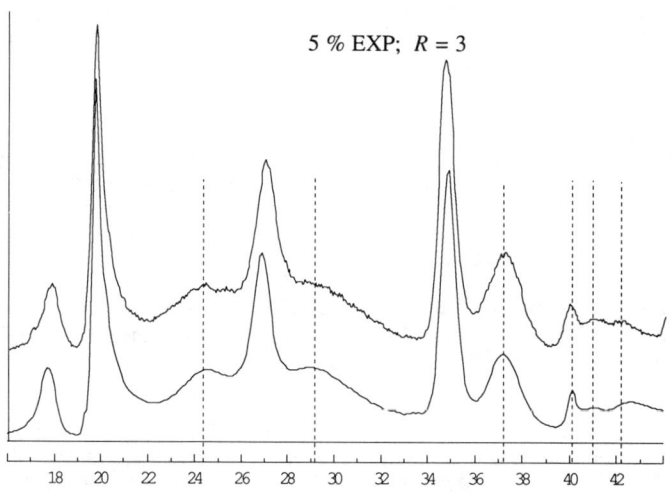

Figure 16. Comparison between calculated (bottom) and experimental diffraction patterns for the Glencoe Shale KB. Non-default parameters are $P_0 = 0.33$, $P_{cv} = 0.35$ and $P_{60} = 0.6$.

Three Dimensional Diffraction from Disordered Illite

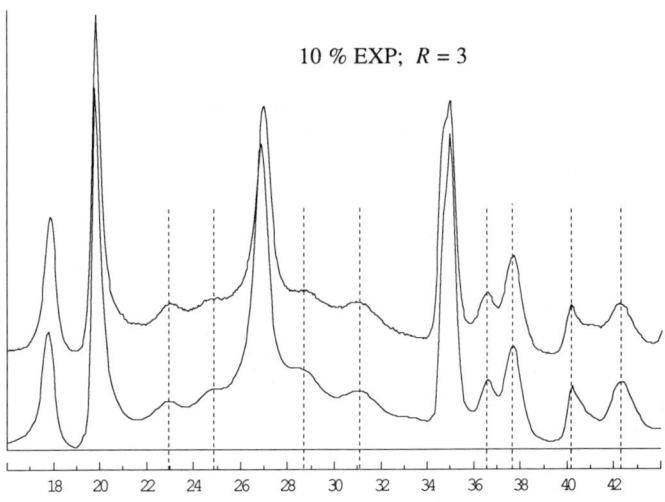

Figure 17. Comparison between calculated (bottom) and experimental diffraction patterns for the Deicke KB. Non-default parameters are $P_0 = 0.65$, $P_{cv} = 0.6$, $P_{60} = 0.2$, $Pf = 0.8$ and $\delta l, Nl = 8, 26$.

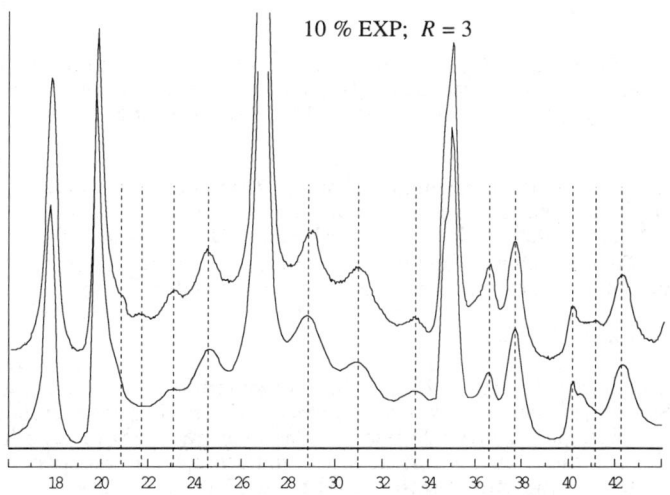

Figure 18. Comparison between calculated (bottom) and experimental diffraction patterns for the Onondaga KB. Non-default parameters are $P_0 = 0.55$, $P_{cv} = 0.8$, and $P_{60} = 0.3$.

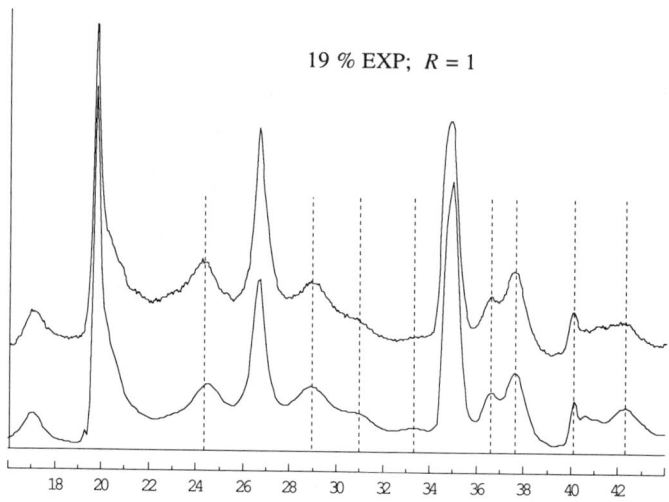

Figure 19. Comparison between calculated (bottom) and experimental diffraction patterns for the Shiphead KB. Non-default parameters are $P_0 = 0.65$, $P_{cv} = 0.4$ and $P_{60} = 0.4$.

The Marias River KB (Figure 20) has a high proportion of interstratified cv layers and, except for % Exp, is similar to the Onondaga KB (Figure 18). The Mancos A KB (Figure 21) is highly disordered ($P_0 = 0.33$) like the Glencoe Shale I/S, but it is disordered mostly by $n.120°$ rotations.

The Utica KB (Figure 22) is interesting because the diffraction pattern clearly shows the dominance of the tv layer type, despite the fact that at 40% Exp, $R = 1$ the crystallites are almost limited to 20 Å-thick illite particles (see Table 1) and might be thought too thin to generate the characteristic tv diffraction pattern. In this case, the breadths of the $\overline{1}12$ and 112 reflections are a consequence of both small crystallite size and rotational disorder.

I/S with High Expandability

The diffraction patterns of Figures 23 and 24 correspond to two K-bentonites with cv structures. The TSI is high because the structures consist of very thin illite crystallites with a dominant proportion of single 2:1 layers (Table 1). The high frequency of particle interfaces causes demodulation of the $20l$; $13l$ diffraction signature toward the two-dimensional bandshape characteristic of turbostratic disorder (c.f. Figure 13).

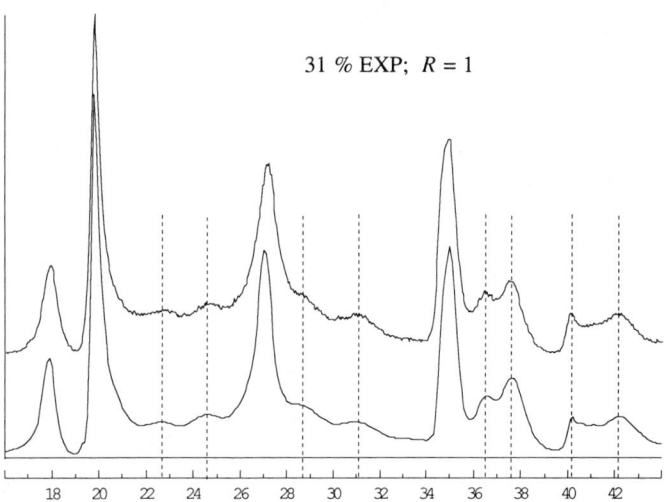

Figure 20. Comparison between calculated (bottom) and experimental diffraction patterns for the Marias River KB. Non-default parameters are $P_0 = 0.75$, $P_{cv} = 0.7$ and $P_{60} = 0.3$.

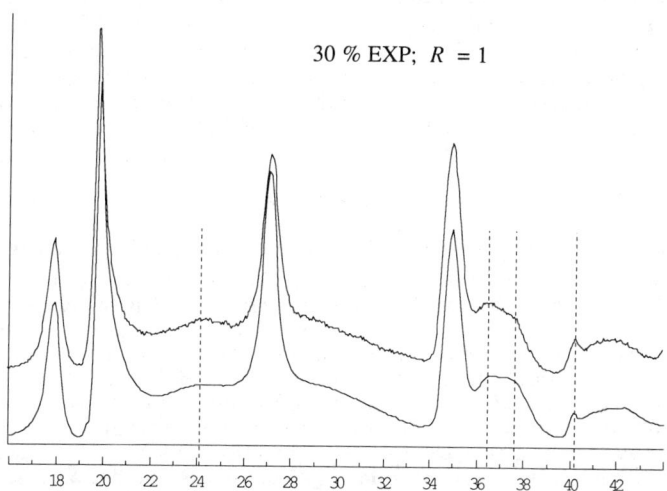

Figure 21. Comparison between calculated (bottom) and experimental diffraction patterns for the Mancos A KB. Non-default parameters are $K = 0.5$, $P_0 = 0.33$, $P_{cv} = 0.3$, $P_{60} = 0.2$ and $N_1 = N_2 = 30$. Muscovite atomic coordinates used (Guthrie and Veblen, 1989).

Reynolds

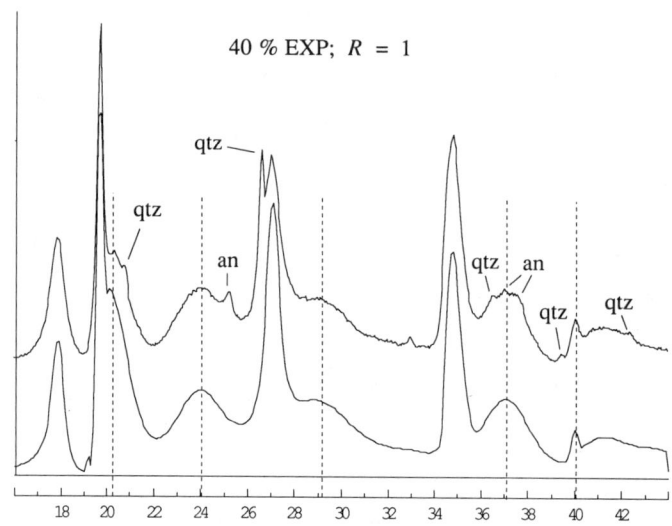

Figure 22. Comparison between calculated (bottom) and experimental diffraction patterns for the Utica Shale KB. Non-default parameters are $K = 0.5$, $P_0 = 0.55$, $P_{cv} = 0$ and $P_{60} = 0.6$. Qtz and an refer to peaks from quartz and anatase.

Diffraction patterns for these two minerals with ethylene glycol solvation were given by Reynolds (1992). The reflection near 22 °2θ was tentatively identified as cristobalite and the peak near 31 °2θ was ascribed to the 004/005. However, the peak near 31 °2θ remained despite dehydration and coincides with the strong $\overline{1}13$ cv reflection; consequently, the interpretation of a cv structure is suggested for these two K-bentonites. The poor agreement between calculated and experimental positions for the illite 002 peak (dehydrated; Figures 23 and 24) is probably due to partial hydration (one water layer) of the samples before or during diffraction analysis.

COMMENTS

The accuracy for P_0 and P_{cv} is believed to be ±0.05 and ±0.1 for P_{60} because such differences make detectable changes in the calculated patterns. The uniqueness of the solutions is more difficult to assess. Interstratified cv-tv structures could perhaps be simulated with a single intermediate unit cell, though attempts to do this with the M layer of Drits et al. (1984) produced poor matches with the experimental patterns. Variable atomic coordinates with respect to % Exp for I/S seems likely because there are differences in tetrahedral Al content over the mixed-layered series. However, the goodness-of-fit demonstrated by Figures 15 to 24 was achieved with invariant atomic coordinates. The diffraction patterns may be more sensitive to the disorder parameters (and perhaps the cv-tv layer type) than they are to the variability of atomic positions over the I/S mixed-layered series. If true, then the model parameters are reliable descriptors for these structures.

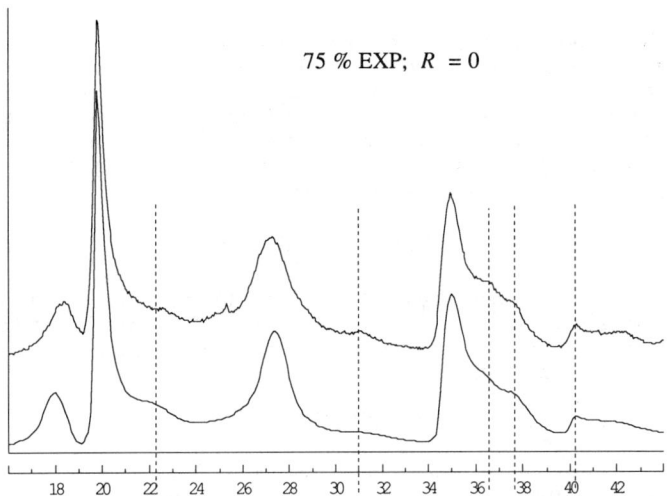

Figure 23. Comparison between calculated (bottom) and experimental diffraction patterns for the Mancos B KB. Non-default parameters are K = 0.4, $P_0 = 0.6$, $P_{cv} = 1$, $P_{60} = 0$, and $\delta l, Nl = 5, 16$.

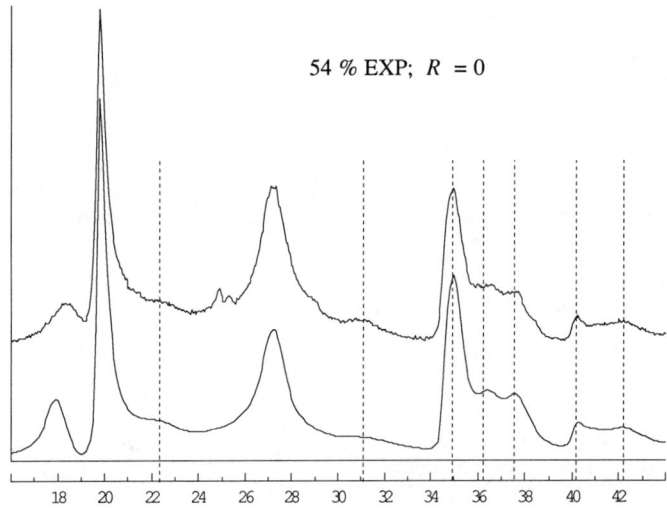

Figure 24. Comparison between calculated (bottom) and experimental diffraction patterns for the Mancos C KB. Non-default parameters are K = 0.3, $P_0 = 1$, $P_{cv} = 1$, $P_{60} = 0$, and $\delta l, Nl = 3, 9$.

The family of illite minerals may be much more complicated than previously thought. In addition to the common and well-described mixed-layered series, there is the variety of structural modifications described here. Many clay mineralogists confine their X-ray diffraction studies to oriented aggregates. Until a suitable data base of high-quality three-dimensional powder patterns of illites are accumulated for a variety of rocks, whatever complexity exists in the clay-size illite family will remain unappreciated, as will the geological significances of crystal structures not accessible by single crystal X-ray diffraction methods.

ACKNOWLEDGEMENTS

Support for this research was provided by American Chemical Society Grant 23613-AC2. In addition, the writer is indebted to S. P. Altaner and D. K. McCarty for providing some of the samples used for study. An early draft of the manuscript benefitted from a thorough review by Steve Guggenheim, however, the author assumes full responsibility for remaining errors.

REFERENCES CITED

Altaner, S. P. and Bethke, C. M. (1988) Interlayer order in illite/smectite: *American Mineralogist.* **73**, 766-774.

Bailey, S. W. (1980) Structures of layer silicates: in *Crystal Structures of Clay Minerals and their X-Ray Identification*, G. W. Brindley and G. Brown, eds., Mineralogical Society, London, 1-123.

Bailey, S. W. (1991) Practical notes concerning the X-ray powder diffraction patterns of clay minerals: *Clays & Clay Minerals* **39**, 184-190.

Bethke, C. M. and R. C. Reynolds (1986) Recursive method for determining frequency factors in interstratified clay diffraction calculations: *Clays & Clay Minerals* **34**, 224-226.

Besson, G., Glaesner, R., and Tchoubar, C. (1983) Le césium, révélateur de structure des smectites: *Clay Minerals* **18**, 11-19.

Brindley, G. W. (1980) Order-disorder in clay mineral structures: in *Crystal Structures of Clay Minerals their X-Ray Identification*, G. W. Brindley and G. Brown, eds., Mineralogical Society, London, 125-195.

Brindley, G. W. and Méring, J. (1951) Diffraction des rayons X par les structures en couches desordonnees: *Acta Crystallographica* **4**, 441-447.

Dollase, W. A. (1986) Correction of intensities for preferred orientation in powder diffractometry; Application of the March model: *Journal of Applied Crystallography* **19**, 267-272.

Drits, V. A., Plançon, B. A., Sakharov, B. A., Besson, G., Tsipursky, S. I. and Tchoubar, C. (1984) Diffraction effects calculated for structural models of K-saturated montmorillonite containing different types of defects: *Clay Minerals* **19**, 541-561.

Drits, V. A., and Tchoubar, C. (1990) *X-Ray Diffraction by Disordered Lamellar Structures*: Springer-Verlag, New York, 371 p.

Guthrie, G. D., and Veblen, D. R. (1989) High-resolution transmission electron microscopy of mixed-layered illite/smectite: Computer simulations: *Clays &Clay Minerals* **37**, 1-11.

Ergun, S. (1970) X-ray scattering by very defective lattices: *Physical Review B*, **131**, 3371-3380.

James, R. W. (1965) *The Optical Principles of the Diffraction of X-Rays*: Cornell University Press, Ithaca, New York, 664 p.

Méring, J. (1975) Smectites: in *Soil Components, Vol 2. Inorganic Components*, J. E. Gieseking, ed., Springer-Verlag, New York, Chp. 4.

Méring, J. and Oberlin, A. (1967) Electron-optical study of smectites: *Clays &Clay Minerals*, **15**, 3-25.

Nadeau, P. H., Tait, J. M., McHardy, W. J. and Wilson, M. J. (1984) Interstratified XRD characteristics of physical mixtures of elementary clay particles: *Clay Minerals* **19**, 67-76.

Plançon, A. (1981) Diffraction by layer structures containing different kinds of layers and stacking faults: *Journal of Applied Crystallography* **14**, 300-304.

Plançon, A. and Tchoubar, C. (1977a) Determination of structural defects in phyllosilicates by X-ray powder diffraction--I. Principle of calculation of the diffraction phenomena: *Clays & Clay Minerals* **25**, 430-435.

Plançon, A. and Tchoubar, C. (1977b) Determination of structural defects in phyllosilicates by X-ray powder diffraction--II. Nature and proportion of defects in natural kaolinite: *Clays & Clay Minerals* **25**, 436-450.

Plançon, A., Besson, G., Gaultier, J. P., Mamy, J., and Tchoubar, C. (1979) Qualitative and quantitative study of structural reorganization in montmorillonite after potassium fixation: *Proceedings of the VIth International Clay Conference, Oxford 1978*, 45-54.

Plançon, A., Giese, R. F., and Snyder, R. (1988) The Hinckley index for kaolinites: *Clay Minerals* **23**, 249-260.

Reynolds, R. C. (1980) Interstratified clay minerals: in *Crystal Structures of Clay Minerals and their X-Ray Identification*, G. W. Brindley and G. Brown, eds., Mineralogical Society, London, 249-303.

Reynolds, R. C. (1985) *NEWMOD©, a Computer Program for the Calculation of Basal X-Ray Diffraction Intensities of Mixed-Layered Clays*. R. C. Reynolds, Hanover, N. H. 03755.

Reynolds, R. C. (1989a) Principles of powder diffraction: in *Modern Powder Diffraction*; Bish, D. L., and Post, J. E. (eds.): *Reviews in Mineralogy* **20**, Mineralogical Society of America, 1-17.

Reynolds, R. C. (1989b) Diffraction by small and disordered crystals: in *Modern Powder Diffraction*; Bish, D. L., and Post, J. E. (eds.): *Reviews in Mineralogy* **20**, Mineralogical Society of America, 145-181.

Reynolds, R. C. (1990) A preliminary study of order/disorder and polytypism in mixed-layered illite/smectite: in *Programs with Abstracts, 27th Annual Meeting, Clay Minerals Society*, Columbia, Missouri, 1983, p. 105 (abstract).

Reynolds, R. C. (1992) X-ray diffraction studies of illite/smectite from rocks, <1 µm randomly oriented powders, and <1 µm oriented powder aggregates: The absence of laboratory-induced artifacts: *Clays & Clay Minerals* **40**, 387-396.

Reynolds, R. C. and Thomson, C. H. (1993) Illite from the Potsdam Sandstone of New York, a probable noncentrosymmetric mica structure: *Clays & Clay Minerals* (In Press).

Sakharov, B. A., Besson, G., Drits, V. A., Kamenava, M. Yu, Salyn, A. L. and Smoliar, B. B. (1990) X-ray study of the nature of stacking faults in the structure of glauconites: *Clay Minerals* **25**, 419-435.

Tsipursky, S. I. and Drits, V. A. (1984) The distribution of octahedral cations in the 2:1 layers of dioctahedral smectites studied by oblique-texture electron diffraction: *Clay Minerals* **19**, 177-193.

Wright, A. C. (1973) A compact representation for atomic scattering factors: *Clays & Clay Minerals* **21**, 489-490.

Yoder, H. S., and Eugster, H. P. (1955) Synthetic and natural muscovites: *Geochimica et. Cosmochimica Acta* **6**, 157-185.

Zvyagin, B. B., Robotnov, V. T., Sidorenko, O. V., and Kotelnikov, D. D. (1985) Unique mica with noncentrosymmetric layers: *Izvestiya Akad. Nauk SSSR, Geol.* **35**, 121-124 (in Russian).

STUDIES OF CLAYS AND CLAY MINERALS USING X-RAY POWDER DIFFRACTION AND THE RIETVELD METHOD

David L. Bish

CONTENTS

Introduction	80
The Rietveld Method	81
Starting Models for Refinement	83
Transmission electron microscopy	83
Distance least-squares modeling	84
Electrostatic energy minimization	84
Disorder In Clays And Clay Minerals	85
Effects of Disorder on Diffraction Patterns	86
Applications Of The Rietveld Method To Clay Minerals	91
Crystal Structure Refinements	91
Dickite	91
Kaolinite	91
Chlorite	93
Partial Structure Solution	94
Hydrogen Atoms in Kaolinite	95
Hydrogen Atoms in Dickite	97
Interlayer Structure of Kaolinite Intercalates	98
Exchangeable Cations and Water in Sepiolite	99
Quantitative Analysis	100
Theory	100
Application	104
Refinement of Unit-Cell Parameters	106
Analysis of Peak Broadening	108
Goethite	110
Sample Refinement	112
Sample Preparation and Data Collection	112
Refinement Strategies	114
Conclusions	116
Acknowledgments	117
References Cited	117

STUDIES OF CLAYS AND CLAY MINERALS USING X-RAY POWDER DIFFRACTION AND THE RIETVELD METHOD

David L. Bish

Earth and Environmental Sciences
Los Alamos National Laboratory
Los Alamos, New Mexico 87545

INTRODUCTION

The Rietveld method was originally developed (Rietveld, 1967, 1969) to refine crystal structures using neutron powder diffraction data. Since then, the method has been increasingly used with X-ray powder diffraction data, and today it is safe to say that this is the most common application of the method. The method has been applied to numerous natural and synthetic materials, most of which do not usually form crystals large enough for study with single-crystal techniques. It is the ability to study the structures of materials for which sufficiently large single crystals do not exist that makes the method so powerful and popular. It would thus appear that the method is ideal for studying clays and clay minerals. In many cases this is true, but the assumptions implicit in the method and the disordered nature of many clay minerals can limit titsapplicability. This chapter will describe the Rietveld method, emphasizing the assumptions important for the study of disordered materials, and it will outline the potential applications of the method to these minerals. These applications include, in addition to the refinement of crystal structures, quantitative analysis of multicomponent mixtures, analysis of peak broadening, partial structure solution, and refinement of unit-cell parameters.

An important requirement with the Rietveld method, and one that is often only *assumed* to be met, is that the diffraction pattern must exhibit only Bragg diffraction effects. Unfortunately, the layered nature of many clay mineral structures makes them prone to stacking disorder, giving rise to two-dimensional diffraction effects. This aspect of their structures makes the application of the Rietveld method to clay minerals difficult and, in many cases, unwarranted. Thus it is critical to understand the limitations of the method when applying it to poorly ordered materials. In particular, it is important to recognize that two-dimensional diffraction bands cannot be simulated simply by broadening the Bragg reflections.

In spite of these limitations, the Rietveld method has been successfully applied to several well-ordered clays, including kaolinite, dickite, nacrite, chlorite, and the chain-structure clay mineral sepiolite. In these cases, samples were specially chosen which showed little or no evidence of two-dimensional diffraction effects. These studies yielded detailed structural information, including positions of exchangeable cations, data on octahedral and tetrahedral cation ordering, and the detailed effects of temperature on the structures. For well-ordered materials such as these, partial structure solution is an important adjunct to Rietveld refinement, usually using difference-Fourier syntheses. The coupled use of these two procedures has been shown to be very effective in locating H atoms (using neutron diffraction data) and interlayer intercalate species such as hydrazine and H_2O molecules in kaolin minerals. Application of the Rietveld method to a chlorite with semi-random stacking (*i.e.*, some two-dimensional diffraction effects) illustrates some of the problems inherent with disordered materials. The refinement yielded precise cell parameters, but atomic positions and occupancies were artifacts due to the $b/3$ layer shifts.

Use of the Rietveld method with non-layer-structure clay minerals has been fruitful, as these materials are not usually hampered by two-dimensional diffraction effects, although they are often very poorly crystalline (*i.e.*, they have very small crystallite sizes). For example, several studies of goethite (FeOOH) have elucidated the detailed structural changes occurring with isomorphous substitutions of Al, Cr, and Mn for Fe; other studies have identified the sources of the pronounced peak broadening which appear to depend partially on these isomorphous substitutions. Reflections in Mn-substituted goethites are predominantly strain broadened, due to Jahn-Teller distorted octahedra, whereas Al-substituted goethites exhibit primarily crystallite-size broadening (Bish and Ebinger, 1989). Even with very broad reflections, the Rietveld method appears to give useful structural information, as long as diffraction data show no non-Bragg diffraction effects.

THE RIETVELD METHOD

Conceptually, the Rietveld method is different from other more traditional techniques for evaluating X-ray powder diffraction data in that a refinement uses digital diffraction data and is done on a point-by-point (step-by-step) basis instead of dealing with individual reflections. Specifically, a Rietveld refinement minimizes R, the sum of the weighted, squared differences between observed and calculated intensities at every 2θ step in a digital powder pattern,

$$R = \sum_i w_i |y_i(o) - y_i(c)|^2, \quad (1)$$

where $y_i(o)$ and $y_i(c)$ are observed and calculated intensities at point i, and w_i is the weight assigned to each intensity. The calculated intensities at each point (or 2θ step), $y_i(c)$, are determined by summing the contributions from background and all neighboring Bragg reflections as:

$$y_i(c) = S \sum_k \left(p_k L_k |F_k| G(\Delta\theta_{ik}) P_k \right) + y_{ib}(c), \quad (2)$$

where S is a phase-specific scale factor, p_k is the multiplicity factor, L_k is the Lorentz and polarization factor for the kth reflection, F_k is the structure factor for an individual reflection for a particular phase, $G(\Delta\theta_{ik})$ is a reflection profile function, θ_{ik} is the Bragg angle for the kth reflection, P_k is a preferred orientation function, and $y_{ib}(c)$ is a refined background.

Bish

Equation (2) clearly illustrates a very important aspect of Rietveld refinements, namely that the method assumes the presence of Bragg diffraction and the absence of non-Bragg diffraction effects. Thus, only integral values of h, k, and l are allowed and the structure factor cannot vary across a reflection. The latter is important with poorly crystalline materials, and the Rietveld method cannot be rigorously applied in these cases because it assumes a given structure factor for each reflection. In addition, because of these factors, samples ideally must not possess any layer-stacking disorder or turbostratic stacking. These aspects of Rietveld refinements are of utmost importance when considering applications to clays and clay minerals, and unfortunately these requirements rule out most clay minerals.

In the absence of two-dimensional diffraction effects, preferred orientation of crystallites is typically the most significant problem encountered in a refinement. Because preferred orientation is such an important consideration when dealing with many clays and clay minerals, it is worthwhile to describe the methods of correcting for this effect. The first popular analytical treatment of preferred orientation in the Rietveld method was the use of the correction factor, G_{hkl} (Rietveld, 1969),

$$G_{hkl} = \exp(-G_1 \alpha^2), \tag{3}$$

where α is the acute angle between the normal to the crystallites and the scattering vector. G_1 is a refinable parameter and is a measure of the half-width of the Gaussian distribution of the normals about the preferred orientation direction. This formulation has met with some success with neutron diffraction data and was slightly modified by Toraya and Marumo (1981) to yield a correction factor (G_{hkl}) of the form

$$G_{hkl} = G_2 + (1 - G_2)\exp(-G_1 \alpha^{2)}), \tag{4}$$

where G_1 and α are analogous to G_1 and α in Eq. (3) and G_2 is the (refinable) fraction of oriented crystallites.

Neither Eq. (3) nor (4) has been very successful with X-ray powder diffraction data, for which preferred orientation is usually a greater problem. In an attempt to rectify this situation, Dollase (1986) examined a variety of analytical forms previously proposed to represent pole-figure profiles. Dollase concluded that the March (1932) function, which describes the pole-density distribution yielded by rigid-body rotation of inequant crystallites upon axially symmetric volume-conserving expansion or compression, has significant advantages over other proposed functions. The procedure used in a Rietveld refinement with the March function correction consists of fitting the pole-density profile using a function with a single variable parameter. The function used by Dollase (1986) is

$$G_{hkl} = \left(r^2 \cos^2 \alpha + r^{-1} \sin^2 \alpha\right)^{-3/2}, \tag{5}$$

where α is as defined for Eq. (3) and *r is* the only adjustable coefficient related to the degree of preferred orientation. Dollase also pointed out that most of the earlier preferred orientation correction functions work well with weakly developed preferred orientation, such as is common with non-platy minerals in a neutron diffraction experiment, but they do not work well with typical materials in an X-ray diffraction experiment. Moreover, an important shortcoming of the previous corrections [*e.g.*, Eq. (3) or (4)] is that none is normalized to unit integral which means that any change in a refinable parameter will be counterbalanced by a change in the overall scale factor. This property of the correction is particularly important for quantitative analysis using the Rietveld method. The author's experience has shown that the March function suggested by Dollase works well for low to high degrees of preferred orientation in platy materials and fairly well for rod-like materials with low degrees of preferred orientation. Many of the popular Rietveld computer programs now include the March function for preferred orientation corrections.

Starting Models for Refinement

By its very nature, structure refinement relies on the existence of a reasonably accurate starting structure model. The very large number of published refined structures available in the literature provides a starting point for most structure refinements. Diffraction data are then used to improve, or to refine, the model to yield optimum agreement between observed and calculated observations. In addition, geometrical constraints and crystal chemical intuition have played a part in solving structures. However, as the complexity of structures increases, additional information is usually required and can often be obtained from transmission electron microscope (TEM) images, distance least-squares (DLS) modeling, and/or energy minimizations (Bish, 1992).

Transmission electron microscopy. Additional structural data have come, in recent years, in the form of high-resolution transmission electron microscopy (HRTEM) images, which have been used to develop structure models (*e.g.*, Banfield et al., 1991). The structure models for several cases were derived from HRTEM images which were then further optimized using DLS modeling to develop useful starting models for Rietveld refinement. Within the past twenty years, transmission electron microscopy has advanced to the point where it can produce images of the atomic arrangement in a crystal, often easily yielding insights into general structural schemes and, in some cases, solving apparently intractable problems (*e.g.*, Veblen, 1985; Buseck and Veblen, 1988). It is important to note, however, that HRTEM images must be used with caution. Minor changes in focus in the microscope can cause large changes in the high-resolution image, greatly changing the images (*e.g.*, Guthrie and Veblen, 1990). In order to interpret the images unambiguously, it is necessary to couple HRTEM observations with modeling of the images using a multi-slice calculation; such calculations require at least an approximate knowledge of the structure (*e.g.*, O'Keefe, 1984; Post and Veblen, 1990).

Distance least-squares modeling. DLS modeling (Meier and Villiger, 1969) is a powerful method for determining whether or not a particular model is geometrically possible. The method is similar in many respects to conventional crystal structure refinement, particularly in that it requires a starting model. These models can be obtained by analogy with other structures or by crystal chemical reasoning (or from HRTEM images). Fortunately, the DLS method is very robust and calculations will usually converge, even given very poor starting models. An optimum distance model is obtained by varying atomic coordinates and unit-cell parameters to minimize discrepancies between interatomic distances calculated for the model structure and prescribed (input) distances that comprise the observations. Input distances are usually obtained from published, well-refined analogous structures and can often be estimated to within several hundredths of an Å. Weights applied to distances are usually related either to Pauling bond strengths or to some empirical measurement, such as stretching force constants; weights based on the latter emphasize shorter distances and those involving highly charged ions much more strongly than the former method. Known unit-cell parameters can be used in many cases to constrain the refinements further, resulting in more reasonable structures. Unit-cell parameters can usually be obtained by indexing an X-ray powder or electron diffraction pattern using well-established techniques when beginning with unknown or poorly known structures.

DLS modeling of a completely unknown structure requires some imagination and adjunct information (such as HRTEM images), but modeling a variant of a known material can be quite easy. For example, if an intercalation reaction in a layered material such as kaolinite, $Al_2Si_2O_5(OH)_4$, gives rise to a larger unit repeat in the direction perpendicular to the layers, DLS modeling can use the parent structure together with the new unit-cell parameters to represent the new structure. Of course, one must either provide unit-cell parameters or use distance observations sufficient to constrain the structure in three dimensions. After any DLS modeling, calculated diffraction patterns usually provide a quick idea of how reasonable the structure model is, easily revealing whether the symmetry and cell are correct and how closely the model structure approximates the real one. This procedure is similar to that used by Banfield et al. (1991) with HRTEM images, namely obtaining a model structure and determining how well observations agree with those calculated based on the model. In some cases, a DLS model structure can also minimize the problems of false-minima in a Rietveld refinement, as appears to be the case for one study of kaolinite (see discussion in Bish and Von Dreele, 1989).

Electrostatic energy minimization. Electrostatic energy minimization is a final method that is proving increasingly useful in supporting determinations of H-atom orientations and extra-framework species positions. This method combines calculation of the Coulomb portion of the lattice energy with some formulation of the short-range repulsive energies to determine the minimum-energy configuration of a particular arrangement of atoms. Guthrie and Bish (1991) applied this method to a study of the H orientations in kaolinite, for which H positions had just recently been determined using Rietveld refinement/difference-Fourier (ΔF) methods and neutron powder diffraction data. It is very interesting that the results of the electrostatic minimizations reproduced both the H positions *and* the general shape of the potential wells within which the H atoms resided. Thus, this method explained and supported the anisotropic displacement parameters of the H atoms observed in the Rietveld refinement. Giese (1982) and Guthrie and Bish (1991) also used these methods to predict the H positions and orientations in layer silicates for which complete structure refinements have not been performed. Catlow et al. (1986) described numerous uses of electrostatic methods, including studying cation distributions in zeolites, and they commented on the possibilities of coupling energy minimizations with structure refinements using diffraction data.

DISORDER IN CLAYS AND CLAY MINERALS

Clay minerals exhibit a wide range of types of order/disorder, some of which affect the applicability of the the Rietveld method (see Brindley, 1980, for a general discussion). Substitutional disorder, such as the occupancy of a crystallographically distinct site by more than one type of atom, is very common in minerals. Examples of such disorder include the tetrahedral Al/Si distribution in $2M_1$ micas and Fe/Al distribution on the octahedral site in goethite. In general, this type of disorder does not affect the diffraction pattern in a way that would limit the applicability of the Rietveld method. However, substitutional disorder can give rise to so-called lattice microstrain, as a result of the occupation of crystallographically "equivalent" sites by significantly different cations (or anions). An example of such an effect are the Mn-substituted goethites, in which octahedra occupied by Mn are significantly more distorted than those occupied by Fe, producing a crystallite containing unit cells of different dimensions.

Strain broadening contrasts with that produced by the presence of very small crystallite sizes in a sample. Reflections are broadened by the presence of small crystallite sizes in proportion to the inverse of the crystallite size in a particular direction related to the reflection. For example, decreasing the packet thickness in a clay mineral will result in broadening of all reflections with a non-zero l Miller index. Reflections with an l Miller index of zero will be unaffected, all other factors remaining the same. Both strain and crystallite-size reflection broadening can be considered in the profile shape model and thus can be modeled explicitly in a Rietveld refinement. The Rietveld code GSAS (Larson and Von Dreele, 1988) can also model *anisotropi*c crystallite size and strain broadening, in which different classes of reflections are broadened differently by both factors.

An additional type of disorder that is very common with clay minerals is layer stacking disorder. This type of disorder has been described in detail by Reynolds (1989) and only a summary will be provided here. There are several types of disorder common with clays and clay minerals, and perhaps the first step in a hypothetical sequence from ordered to disordered materials is so-called semi-random stacking. Semi-random stacking is seen most often with chlorites, in which the 2:1 layers above and below the interlayer hydroxyl sheet maintain hydrogen bond contact by adopting positions related by $\pm b/3$. At each interface above and below the interlayer sheet, three positions are available, and the layers on either side of the interface adopt one of these positions randomly. Because the layers "choose" randomly from among a limited number of choices, the resulting structures are termed "semi-random." The layer shifts are all related by $\pm b/3$ irrespective of which position is chosen. Thus, reflections with indices in which $k \neq 3n$ will be streaked, producing a two-dimensional diffraction band.

If individual layers or packets of layers are randomly rotated or displaced in the X-Y plane but the repeat is maintained along Z, a turbostratic structure results. Such a structure is much like a deck of playing cards that has been randomly rotated and displaced so that the cards are not lined up. In this case, the crystallites scatter as single layers and the only dimensional repeat is along Z. Therefore, only 00l Bragg reflections will occur, and the remaining hkl reflections will be smeared out into two-dimensional diffraction bands. In reciprocal space, the reciprocal lattice nodes that occur for an ordered crystal degenerate into a set of rods along $c*$. The intensity varies continuously along these rods, and the breadth or circumference of the rods depends upon the crystal perfection in the X-Y plane (Figure 1). Reynolds (1989) described several methods of calculating the intensity distribution diffracted from a powder of such material, and he showed that such calculations depend on the type of layer stacking defects (simple rotations, shifts, and/or layer separations) and the statistical distribution of such defects. Clearly, such a calculation is not reproduced by the mathematics in Eq. (2), and this equation does not in any way model two-dimensional diffraction effects. Figure 2, comparing observed kaolinite and halloysite diffraction patterns with calculated kaolinite patterns broadened using the profile shape function, further reinforces this concept and shows that simple broadening in no way models the two-dimensional diffraction effects in the diffraction pattern of halloysite.

A final step toward disordered materials is a material possessing little or no long-range three-dimensional order, that is, an amorphous or glassy material. Reynolds (1989) presented a brief overview of diffraction from such materials; suffice it to say that, as with two-dimensional diffraction effects, diffraction effects from an amorphous material, cannot be modeled by the Rietveld method although some programs have the capability to model the "amorphous humps" as background.

Effects of Disorder on Diffraction Patterns

The effects of isotropic or anisotropic crystallite size and/or strain broadening are to broaden some or all reflections in a diffraction pattern, and these effects can be reasonably well modeled and distinguished with the Rietveld method. Bish and Ebinger (1989) showed for goethite that it can be fallacious to assume that broad reflections are always due to small crystallite size effects, as is implicitly assumed when applying the Scherrer equation. Strain broadening effects are not often identified in clays and clay minerals but they are probably more common than realized, as a result of substitutional and displacive defects.

Virtually all layer silicates exhibit at least some layer stacking disorder due to relatively weak forces that hold the layers together. The results of Walker and Bish (1992) for a IIb Mg-chamosite show this effect and also illustrate some of the pitfalls of attempting a structure refinement using data exhibiting some non-Bragg diffraction effects (Figure 3). Note in Figure 3b that the calculated pattern represents the observed pattern reasonably well, but in Figure 3a, there is very poor agreement between observed and calculated data, and only a broad elevation of background is present with no discrete peaks. Obviously, attempting to refine a crystal structure using such data will result in an artifact, due to the absence of considerable diffraction information (*i.e.*, discrete $k \neq 3n$ reflections).

Studies of Clays Using the Reitveld Method

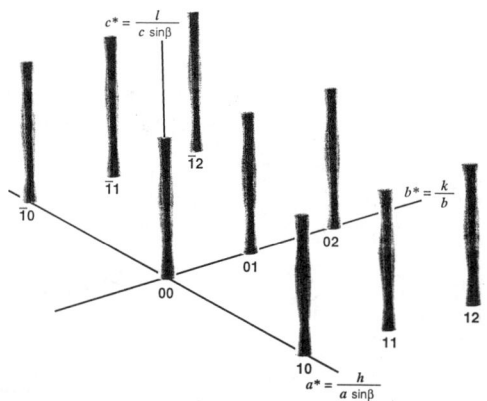

Figure 1. Schematic model of the reciprocal lattice of a turbostratically stacked layer silicate. Note that the reciprocal lattice nodes have degenerated into rods along c^* and the intensity along these rods varies as a function of l.

Turbostratic stacking, or the absence of any significant long-range order in the Z direction, occurs in essentially all smectites, most vermiculites, and many other 2:1 and 1:1 layer silicates. As described above, this type of stacking results in the loss of most or all three-dimensional diffraction information, and diffraction patterns consist of broadened $00l$ reflections and two-dimensional diffraction bands. Figure 4 shows the features expected from a turbostratically stacked layer silicate, in this case Na-SWy-1 smectite, with broadened $00l$ reflections and distinct two-dimensional diffraction bands. Figure 5 a and b, showing unground and ground dickite respectively, illustrate further the difficulties encountered with two-dimensional diffraction bands. Note that the observed data in Figure 5a can be modeled well, but the data in Figure 4 and Figure 5b cannot be modeled correctly or completely using the Rietveld method, because there is no way of correctly modeling the broad, asymmetric bands. However, it is possible to perform a one-dimensional refinement, using only the $00l$ data and refining only those parameters that define the structure in the Z direction (e.g., c unit-cell parameter, z atom parameters, profile parameters for the $00l$ reflections). A significant difficulty with this methodology would be correctly subtracting the intensity contribution of two-dimensional diffraction bands to the overlapping $00l$ reflections at higher angles unless a highly oriented sample were used. The application of the Rietveld method to these minerals will be severely limited because most clays exhibit at least some layer-stacking disorder that gives rise to two-dimensional diffraction effects. However, it is important to emphasize that the application of *any* diffraction method to disordered materials is difficult. For example, there are numerous cases in the literature where a partially disordered crystal was examined using single-crystal methods, and a conventional least-squares refinement was conducted using diffraction data collected from the crystal. It is common in such cases to obtain data at the theoretical reciprocal lattice nodes, correcting for the continuous or semi-continuous background around the nodes, and subsequently to ignore the diffuse scattering between reciprocal lattice nodes. This method is no more correct than the application of the Rietveld method to a diffraction pattern exhibiting two-dimensional diffraction effects.

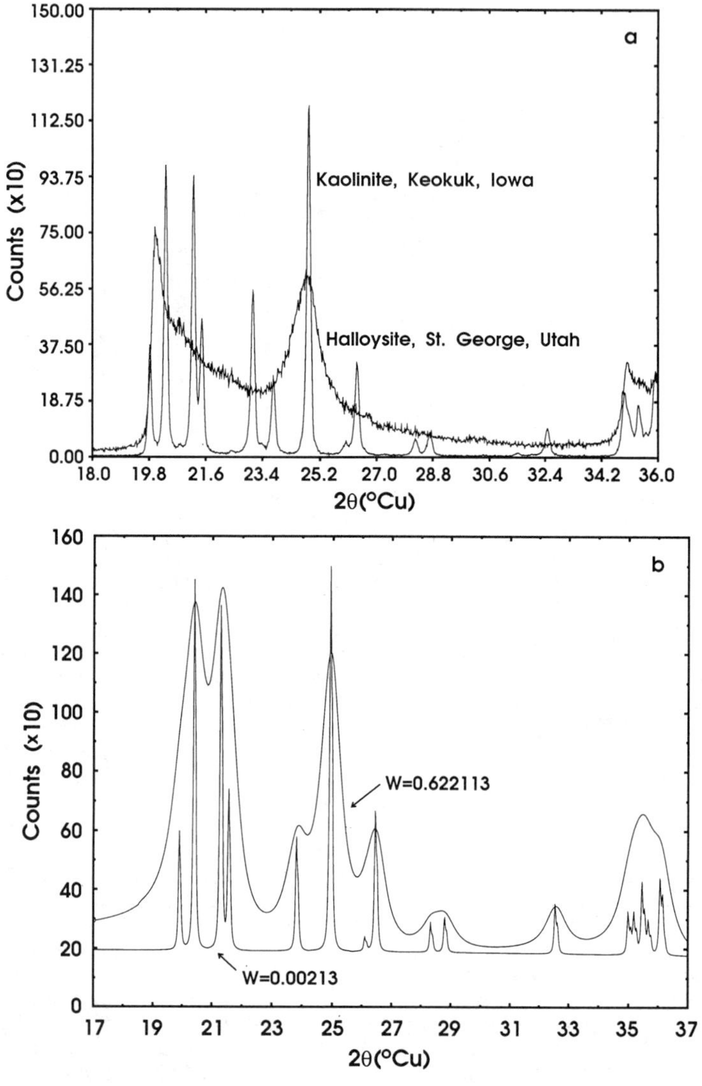

Figure 2. a) Observed patterns (CuKα) of kaolinite and halloysite showing the presence of *hkl* reflections in the kaolinite pattern and two-dimensional diffraction bands in the halloysite pattern; and b) calculated kaolinite patterns broadened using two different W values in the Caglioti *et al.* (1958) profile shape function.

Studies of Clays Using the Reitveld Method

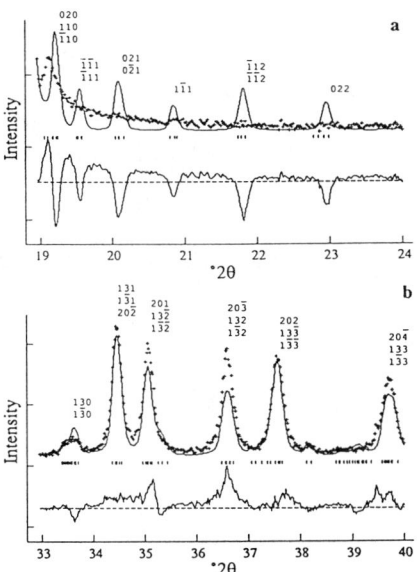

Figure 3. Portions of the observed (+) and calculated (solid line) diffraction patterns for a Mg-chamosite; (a) data from 19 to 24 °2θ (b) data from 33 to 40 °2θ. Solid curve at the bottom of each pattern represents the difference between observed and calculated patterns, and small solid bars beneath observed data represent the calculated positions of all CuKα1 and α2 reflections (from Walker and Bish, 1992).

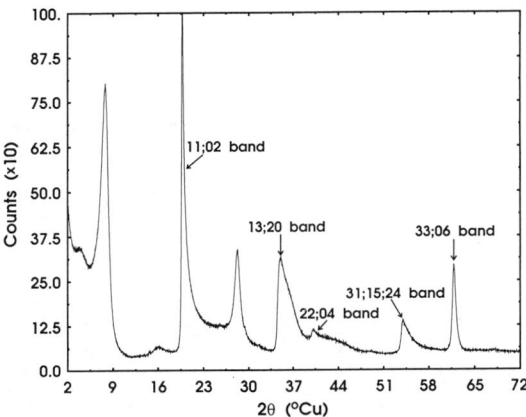

Figure 4. Observed diffraction pattern for a spray-dried Na-SWy-1 smectite. Note the sequence of 00l reflections and the pronounced two-dimensional diffraction bands.

89

Bish

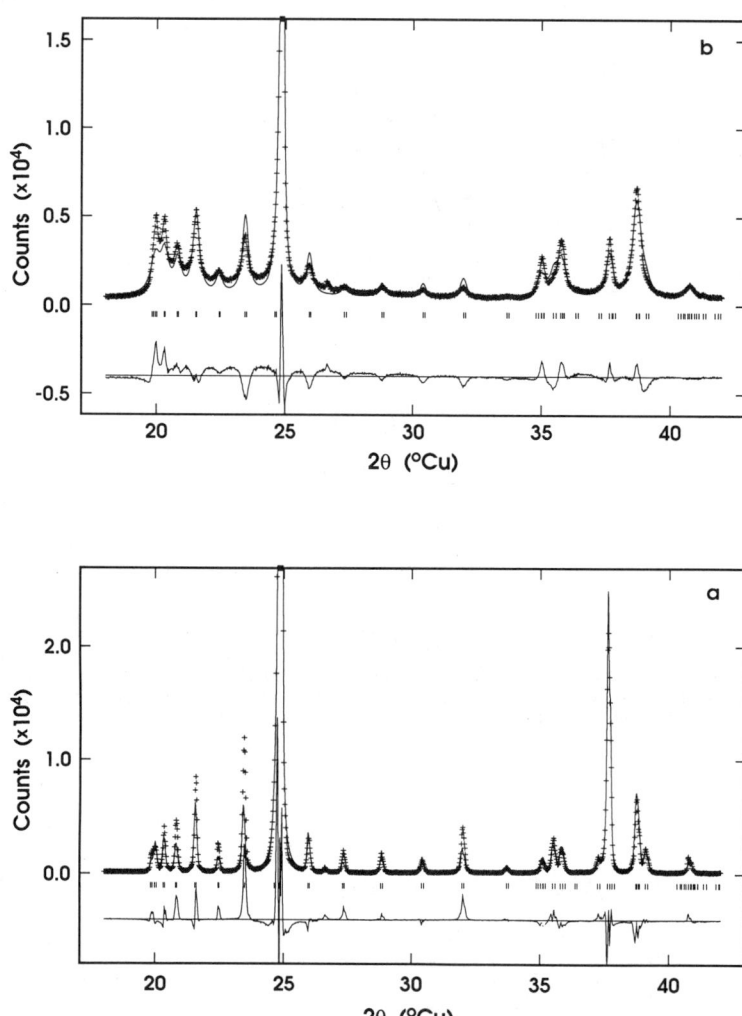

Figure 5. Observed (+) and calculated (solid-line) X-ray diffraction patterns (CuKα) for a) unground dickite, in which the largest discrepancy is for the reflection just above 24 °2θ, due to preferred orientation of the dickite crystallites; and b) dickite that has been ground dry for one hour. Note the enhanced "background" from ~19 to 28 °2θ, due to the presence of grinding-induced layer-stacking disorder, and the inability of the calculated pattern to reproduce this elevation in "background". Tic marks indicate positions for allowed CuKα1 and Kα2 reflections of dickite. Lower curve in each is the difference between observed and calculated profiles.

APPLICATIONS OF THE RIETVELD METHOD TO CLAY MINERALS

Crystal Structure Refinements

Dickite. The first three-dimensional refinement of a clay mineral (dickite) was not published until 1981 (Adams and Hewat, 1981), although the Rietveld method was first devised in 1967. Even then, the application was not well developed for use with X-ray powder diffraction data, so Adams and Hewat used constant-wavelength neutron powder diffraction data in their study. Using the dickite structure of Newnham (1961) (no hydrogen atoms) as a starting model, they were able to locate the hydrogen atoms using a combination of three-dimensional difference-Fourier syntheses and one-dimensional Fourier syntheses. The final Rietveld refinement, incorporating all four H's, yielded an arrangement for the non-H atoms not significantly different from that determined by Newnham (1961). However, the H positions were significantly different from those determined by Giese and Datta (1973) using electrostatic modeling. These differences were ascribed to difficulties in accurately modeling H in the electrostatic calculations. However, Adams and Hewat perceptively commented that further advances in the application of the Rietveld method to the study of clays and clay minerals would require an improved technique for accommodating anisotropic broadening of reflections and some way of incorporating the types of disorder leading to two-dimensional diffraction effects. As will be seen below, the former has been accomplished but the latter remains a major stumbling block.

Bish and Johnston (1993) re-examined dickite using low-temperature neutron powder diffraction data, the Rietveld method and difference-Fourier syntheses, and Fourier-transform infrared spectroscopy. Their refined non-hydrogen structure was essentially identical to other published structures, but their hydrogen positions were distinct. They were able to uniquely assign all four O-H stretching vibrations in infrared spectra to individual OH groups in the dickite structure, and they provided a structural basis for the temperature dependence of all four O-H stretching vibrations. Contrary to published low-temperature infrared spectra, there is no evidence that dickite possesses symmetry lower than Cc at low temperatures.

Kaolinite. One of the first Rietveld refinements of a clay mineral using X-ray powder diffraction data was published by Suitch and Young (1983) who refined the structure of Keokuk, Iowa, kaolinite assuming space group $P1$. Concurrently, Adams (1983) reported the results of a Rietveld refinement of St. Austell, Cornwall, kaolinite in space group $C1$ done with neutron powder diffraction data. These refinements have been discussed at length by Bish (1993).

The most recent X-ray Rietveld study of kaolinite was by Bish and Von Dreele (1989), who used CuKα X-ray powder diffraction data to refine the non-hydrogen positions in the structure. Data were collected in two ranges, from 10° to 90 °2θ counting for 4.0 s every 0.02 °2θ and from 80° to 150 °2θ counting for 12.0 s every 0.02 °2θ. This data-collection strategy is significantly different from that used in other Rietveld refinements in two ways. First, data are typically collected only to ~100 °2θ because of the incorrect impression that high-angle reflections are so overlapped that they are not useful. In addition, collecting data in multiple ranges provides high-angle data with precision comparable to that for the more intense low-angle diffraction intensities.

Bish

Bish and Von Dreele (1989) initially used the Suitch and Young (1983) results as a starting model. However, in spite of attempts to influence the refinement, including incorporating a large preferred orientation correction and using heavily-weighted soft distance constraints (the practice of including predicted interatomic distances as weighted observations, similar to a DLS refinement), their refinement converged to a result very similar to that obtained by Suitch and Young. It was only after using a DLS structure as a starting model that the refinement converged to a final structure with reasonable bond lengths and a significantly lower R factor. These results strongly suggested that Suitch and Young (1983) and Young and Hewat (1988) obtained a false-minimum structure for kaolinite. The refinement of Bish and Von Dreele (1989) also illustrated another benefit of Rietveld refinement. After refining the kaolinite structure using X-ray data for Keokuk kaolinite, systematic differences were seen between the observed and calculated data plots (Figure 6), particularly at ~20.9, 22.5, and 23.5 °2θ These differences were eventually attributed to the presence of ~4% dickite that had been missed in the diffraction patterns by all previous investigators. Additional phases or errors in the model are very apparent because the Rietveld method fits a calculated pattern based on the model structure to the observed data.

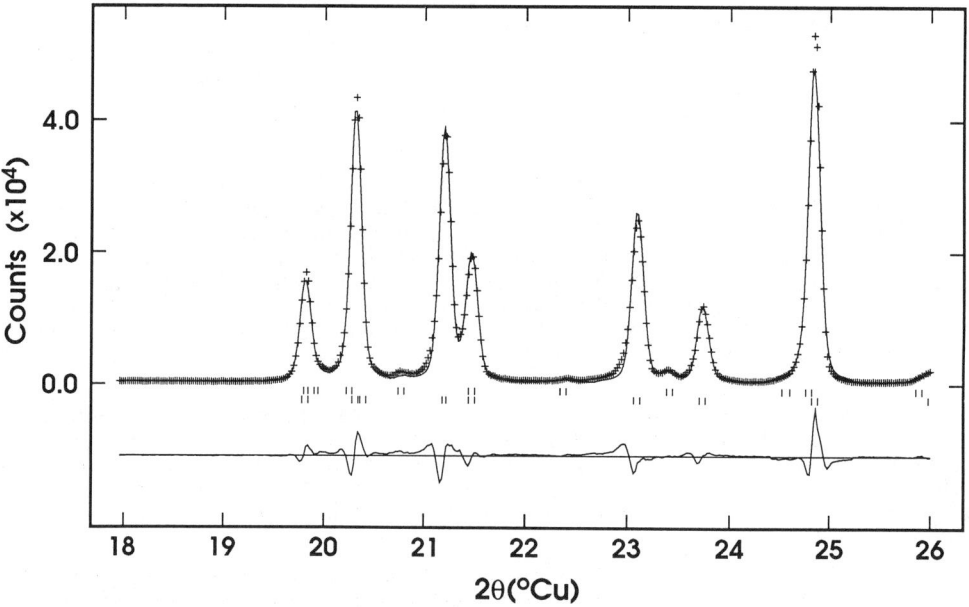

Figure 6. Observed (pluses) and calculated (solid-line) X-ray diffraction pattern for Keokuk kaolinite from 18.0° to 26.0 °2θ, showing the presence of several dickite reflections. Tic marks indicate positions for allowed CuKα1 and Kα2 reflections of dickite (upper) and kaolinite (lower). Small offset between observed and calculated reflection positions is due to a refined sample-displacement correction, some of which is probably due to specimen transparency. Lower curve is the difference between observed and calculated profiles (modified from Bish and Von Dreele, 1989).

The low- and high-temperature structure refinements of Bish (1989) represent a further application of the Rietveld method to kaolinite. He obtained X-ray powder diffraction data at 88, 294, and 573 K and performed complete structure refinements. From 88 to 573 K, the a and b unit-cell parameters increased by only 0.05% and 0.01%, respectively, whereas the c parameter increased by 0.42%. The total volume increase was 0.59%. The markedly anisotropic thermal expansion is consistent with results obtained for other layer silicates. The atomic position results were of relatively low precision, but the cause of the anisotropic thermal expansion *appeared* to be a lengthening of octahedral Al-(O,OH) bonds, resulting in a thickening of the octahedral sheet. These results are not in agreement with limited data for other layer silicates, in which the source of anisotropic expansion appears to be an increase in the interlayer separation. In fact, a recent low-temperature Rietveld refinement of the kaolinite structure (Bish, 1994) confirms that the larger thermal expansion along the Z direction is due to an increase in the interlayer separation.

Chlorite. Although many of the applications of the Rietveld method in clay mineralogy have been to minerals of the kaolin group, a few other clays and clay minerals have been studied using this method. Crystal structures of various chlorite polytypes have been determined primarily using single-crystal X-ray diffraction methods. Many chlorite samples exhibit random or semi-random stacking, and regular stacking sequences are relatively rare. Typical single-crystal studies of chlorites involve examination of many single crystals until a crystal with a regular stacking sequence is discovered (see Bailey, 1988, for a discussion of chlorite stacking sequences). Recently, Walker and Bish (1989), Rakovan and Guggenheim (1991), and Walker and Bish (1992) applied Rietveld refinement methods to several IIb chlorite powders (samples filed, not ground), including two trioctahedral IIb chlorites (chamosite and clinochlore) and a IIb di,trioctahedral chlorite (sudoite). Both the chamosite and sudoite samples studied by Walker and Bish were somewhat disordered as judged by their X-ray powder diffraction patterns. The chamosite refinement, in space group $C\bar{1}$, yielded a final R_{wp} of 15.7% and gave very precise unit-cell parameters. This R_{wp} value is typical of those obtained with complex, poorly ordered materials; average R_{wp} values for Rietveld refinements are larger and are not directly comparable to conventional R values for single-crystal structure refinements. The final fit was good for $k = 3n$ reflections but was very poor for $k \neq 3n$ reflections due to the presence of semi-random stacking in the sample (Figure 3). Although refinement of atomic positions for chamosite was not successful, site-occupancy refinements for the octahedral sites consistent with observed chemistry indicated that Fe and Mg were equally distributed between M1, M2, and M3, whereas M4 was occupied exclusively by Al. The March function preferred orientation correction for chamosite was 0.66, which indicates a large degree of preferred orientation even though the sample was filed and back mounted. Although the final refined structure *appeared* reasonable because most of the atoms repeat at intervals of about $b/3$, it was probably largely an artifact.

Results for the sudoite sample were obviously inferior to those for clinochlore even though the final R_{wp} for sudoite was lower (12.9%). Refinement of atomic positions was again unsuccessful, and site-occupancy refinements suggested the presence of octahedral vacancies that were unsupported by compositional data. It thus appears that the occupancy data may be an artifact of the refinement. Preferred orientation was less for this sample (correction = 0.73), due to the fine-grained nature of the powder. In spite of the difficulties in site-occupancy and atomic position refinements, precise unit-cell parameters were obtained, although they are of questionable accuracy due to the lack of complete three-dimensional information. The clinochlore refinement of Rakovan and Guggenheim (1991) likewise exhibited evidence of systematic errors, attributed by the authors to the presence of a two-layer II*b* polytype intergrown with the dominant one-layer II*b* polytype.

These chlorite refinements are good examples of the limits in applying Rietveld refinement methods to clay minerals. In spite of the fact that the refinements progressed normally and smoothly and yielded very precise unit-cell parameters and low R_{wp} values, the refined structural information appears to be largely invalid. Problems with the refinements are most likely due to the presence of semi-random and random stacking sequences and to the existence of more than one polytype in the chlorite samples. Difficulties arise when attempting to determine structural information from non-Bragg reflections. However, even in the absence of regular stacking sequences, important information can be easily obtained from some chlorites, including precise unit-cell parameters and the relative distribution of heavy (*e.g.*, Fe, Mn, Ni) versus light (*e.g.*, Mg, Al) atoms in octahedral sheets, although compositional data should be used to constrain the results of site-occupancy refinements. Better-ordered chlorites may yield reasonable and useful atomic positions if soft distance constraints consistent with known chemistry are used, although it appears probable that suitably ordered materials are very rare.

Partial Structure Solution

Difference-Fourier (ΔF) synthesis is a method used for refining crystal structures and is particularly well suited for locating scattering density missing from the structure model, such as interlayer species or hydrogen atoms in clay minerals. A ΔF synthesis is unaffected by series termination errors that occur when using limited amounts of data, and this method uses structure factors (F_{obs}) obtained from observed intensities and those calculated using the model structure (F_{calc}). It involves calculating the difference electron density function

$$\Delta F(x,y,z) = \frac{1}{V} \sum_h \sum_k \sum_l \left(F_{obs}^{hkl} - F_{calc}^{hkl} \right) \cos 2\pi (hx + ky + lz). \tag{6}$$

In this equation, F_{calc}^{hkl} represents the structure factors calculated using the assumed structure model and F_{obs}^{hkl} represents the observed structure factors with the same sign as F_{calc}^{hkl}. There are a number of factors to consider when calculating ΔF maps, whether from powder or single-crystal diffraction data. First, in this calculation, the sign (or phase) of F_{obs} is assumed to be the same as F_{calc}. Therefore, if a significant amount of scattering density is missing or misplaced in the model structure, the calculated signs (phases) may be incorrect. For example, Rietveld refinements and ΔF calculations of synthetic Cs A and Y zeolites (Bish, unpublished) were fraught with difficulties because the omission of Cs atoms resulted in many incorrect structure factors (magnitudes and phases). Further, during the initial stages of refinement when some atoms are omitted, it is likely that temperature factors and scale factors will be incorrect, often by a large amount. An additional important consideration when calculating ΔF maps during Rietveld refinement is that the decomposition of overlapping peaks is performed assuming the ratios of the calculated (*i.e.*, model-derived) intensities of the contributing reflections. Thus, results from a set of Rietveld refinements and ΔF-map calculations may be very different from those that would be obtained from profile refinement, where observed intensities are generally model independent. In general, electron-density (or neutron-density) maxima are lower for Rietveld-derived ΔF maps than for those obtained from single-crystal data.

Location of missing atoms is always an iterative process because of these complications. Atoms are found, added to the model if reasonable (in scattering power and position), and another ΔF map is calculated after extraction of observed F's. As more of the correct scattering material is added to the model, the calculated F's (magnitudes and phases) and scale factors become closer to the correct values, and the decomposition of overlapping peaks is improved. These complications also dictate how a Rietveld refinement must be performed in order to obtain reasonable calculated F's. Typically, few parameters are varied, and structural parameters are usually not refined. If structural parameters are varied, they may adjust in an attempt to compensate for the missing scattering power, yielding an unreasonable structure and complicating the interpretation of the ΔF map. In a Rietveld refinement/ΔF map calculation, the scale factor(s) and background parameters are usually varied first, followed by the unit-cell parameters. Some profile parameters may be refined, although heavily overlapped patterns may require either fixing or manual adjustment of these parameters if observed and calculated patterns are significantly different. In practice, it appears that ΔF maps are relatively insensitive to a preferred orientation correction, although it is preferable to use data unaffected by orientation.

Hydrogen Atoms in Kaolinite. One of the more interesting aspects of structural studies of layer silicates and clay minerals is the location of the H atoms that so significantly affect their properties. The structure of kaolinite, $Al_2Si_2O_5(OH)_4$, was outlined as early as 1930, but it was not until 1983 that determinations of the locations of the H positions were first made using neutron diffraction data (Adams, 1983; Suitch and Young, 1983). Previous to 1983, Giese and Datta (1973) and Giese (1982) had modeled H positions using electrostatic modeling methods. Both Adams (1983) and Suitch and Young (1983) used published structures as starting models. Adams determined the H positions using ΔF maps, but his results appear to have been affected by disorder and two-dimensional diffraction effects in the sample. Suitch and Young *assumed* H positions for their starting model and they assumed a lower space group symmetry ($P1$) than accepted for kaolinite ($C1$), resulting in an apparent false-minimum structure. More recently, Young and Hewat (1988) re-refined the kaolinite structure with new neutron powder diffraction data, assuming a complete structural starting model and the lower space group symmetry; they obtained a structure similar to that obtained by Suitch and Young.

In 1990, Bish and Von Dreele re-examined the kaolinite structure using constant-wavelength neutron powder diffraction data and the non-hydrogen structure of Bish and Von Dreele (1989) as a starting model to reconcile the differences between the earlier refinements. It is important that no prior assumptions were made concerning the positions of H in kaolinite. They calculated ΔF maps, and the four largest negative regions of neutron density appeared reasonable for the four H positions (Figure 7). They then refined the complete structure, using ΔF maps and anisotropic refinements to examine the nature of the H sites. There was no significant anisotropy in any of the four H positions, although the inner-hydroxyl H appeared to be slightly smeared along the z direction and the three inner-surface hydroxyl H atoms had larger displacement parameters within the plane of the layers. These results do not agree with those of Adams (1983), Suitch and Young (1983), or Young and Hewat (1988), and they do not support an ordered configuration of the H

Bish

atoms that would require lowering the space group symmetry to *P1*, as suggested by Suitch and Young (1983) and Young and Hewat (1988). It is noteworthy that Guthrie and Bish (1991) reproduced the refined H positions using an electrostatic minimization procedure. Calculated energy contours also showed that the inner-surface hydroxyl H atoms are situated in disk-shaped minima approximately parallel to (001) and the inner-hydroxyl H minimum is approximately perpendicular to (001), in agreement with the results of the Rietveld refinement.

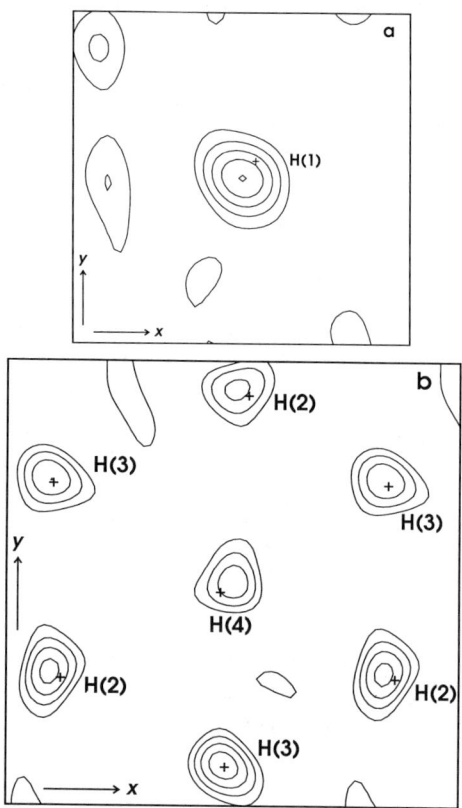

Figure 7. Difference-Fourier maps obtained using neutron powder diffraction data for kaolinite a) in the region of the inner-OH atom, $z = 0.293$, x horizontal, y vertical; b) in the interlayer region at $z = 0.751$, x horizontal, y vertical. Contours are drawn at -0.25, -0.20, -0.15, -0.10, and -0.05 for both maps, and plus (+) symbols represent the final refined positions for each H atom.

Studies of Clays Using the Reitveld Method

Hydrogen Atoms in Dickite. Until recently, there was a similar lack of information on H positions in dickite, a polymorph of $Al_2Si_2O_5(OH)_4$. In probably the first application of the Rietveld method to a layer silicate, Adams and Hewat (1981) examined the dickite structure using Fourier methods and neutron powder diffraction data. Their H positions, obtained on a sample exhibiting some two-dimensional diffraction effects, do not agree well with subsequent single-crystal X-ray determinations (Sen Gupta *et al.*, 1984; Joswig and Drits, 1986). Therefore, Bish and Johnston (1993) re-investigated the structure of dickite using low-temperature time-of-flight neutron powder diffraction data, the Rietveld method, and ΔF syntheses, in addition to variable-temperature Fourier-transform infrared (FTIR) spectroscopy. Using the dickite structure (obtained using single-crystal X-ray diffraction methods) of Joswig and Drits (1986) as a starting model, with no H atoms included, the four H atoms were unambiguously located as the four largest negative regions of density on ΔF maps. Figure 8 shows the three inner-surface hydroxyl H atoms. The minor anisotropy noted for the H atoms in kaolinite was not observed for dickite. The observed H positions were most similar to the refinement of Joswig and Drits (1986), although the O-H bond distances found by Bish and Johnston appear to be more accurate. The OH geometries, and the changes occurring from room temperature to 12 K, are consistent with the FTIR spectra as a function of temperature and lend credence to the Rietveld-refined results. The relative ease with which the H positions were obtained in a powdered clay mineral such as dickite is remarkable when one considers that barely twenty-five years ago such information would have been out of reach using powdered samples.

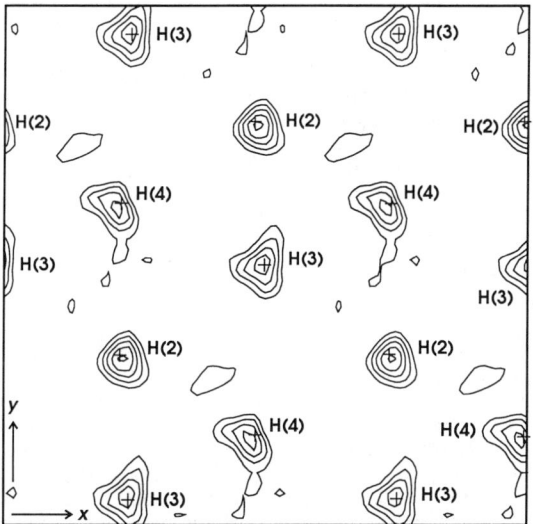

Figure 8. Difference-Fourier map for dickite in the interlayer region (z = 0.352, x horizontal, y vertical) showing the positions of the inner-surface hydrogens. Contours are drawn at -0.5, -0.4, -0.3, -0.2, and -0.1, and plus (+) symbols represent the final refined positions for each hydrogen atom.

It is important to make one final comment regarding the use of ΔF methods to locate H using neutron powder diffraction data. It is common practice in such experiments to attempt to replace H with D to reduce incoherent scattering. If the replacement can be made to go to completion, it should facilitate improved analysis due to a lowered background and improved signal-to-noise ratio. However, one can easily envision a situation in which only partial substitution of D for H is accomplished on a statistical basis. Such a result could produce ΔF maps revealing little or no information about the positions of H or D due to averaging of the negative neutron scattering length of H and the positive neutron scattering length of D. One could create a material with an average neutron scattering length of zero in the H sites.

Interlayer Structure of Kaolinite Intercalates. The determination of the interlayer structure of the kaolinite-hydrazine intercalate is a good example of the utility of combined DLS modeling, ΔF methods, and Rietveld refinement. Although kaolinite does not normally expand, it interacts strongly with several liquids to create expanded intercalates. When kaolinite is exposed to anhydrous hydrazine, N_2H_4, it expands to a material with either a 9.5- or a 10.4-Å basal spacing, depending on the vapor pressure of water. Although the starting kaolinite may not possess good three-dimensional order, the 9.5-Å complex often exhibits three-dimensional order sufficient to apply the Rietveld method. As a first step in the study of the 9.5-Å intercalate, the structure was modeled using DLS, assuming the structure and a, b, α, and γ unit-cell parameters of 7-Å, non-expanded kaolinite, and $\beta=114°$, yielding a value for c of 10.5Å which was obtained from a measured $d(001)$ of 9.52Å. The assumption implicit in the DLS model was that the stacking of layers in the intercalate was the same as in unexpanded kaolinite. The DLS structure was used as the starting model for Rietveld refinement with X-ray powder diffraction data, with the N_2H_4 molecule omitted. The two largest peaks in a ΔF map were in the interlayer region and were 1.40Å apart, consistent with a N_2H_4 molecule. The molecule appears to be canted within the interlayer, and there is evidence from the ΔF maps for positional disorder of the N atom closest to the hydroxyl interlayer surface (Figure 9). Inclusion of these two N atoms in the structure model and continued isotropic refinement yielded an R_{wp} of 25%, compared with 36% without the N atoms, although the refinement is still preliminary. This structure determination/refinement resolved the question of the exact orientation of the N_2H_4 molecule within the kaolinite interlayer, and the refinement result suggests that the intercalate has the same layer stacking as kaolinite. Given the low atomic number of the N and the poor quality of the material, it is remarkable that such a result was obtained with X-ray data.

Comparable results were obtained using this methodology with a similar kaolinite intercalate. Costanzo et al. (1980) created a stable 8.4-Å hydrate of kaolinite by expanding the material with dimethylsulfoxide and then reacting the expanded material in a solution containing NH_4F, thereby partially or completely fluorinating the interlayer hydroxyl groups. The resultant material has an enlarged c repeat and three-dimensional order sufficient to apply the Rietveld method. The starting structure model was obtained using DLS methods, and the location of interlayer water molecules was determined from Rietveld refinement, X-ray powder diffraction data, and ΔF maps (Bish et al., 1992). The largest peak on the ΔF maps was midway between the layers, with only two other minor peaks above zero density. Rietveld refinement, incorporating an O atom at the position of the largest peak, yielded an H_2O occupancy of 0.75, positioned almost exactly midway between the silicate layers. Subsequent ΔF maps revealed no significant areas of electron density.

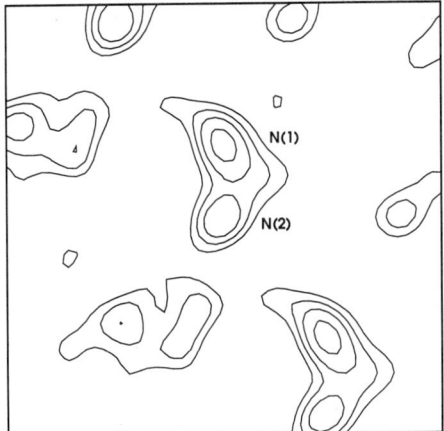

Figure 9. Difference-Fourier map for the kaolinite-hydrazine intercalate in the interlayer region ($z = 0.649$), showing the positions of the two interlayer nitrogen atoms of the hydrazine molecule. Contours are drawn at 0.15, 0.30, 0.45, and 0.65.

Exchangeable Cations and Water in Sepiolite. Sepiolite would appear to be an ideal candidate for the use of Rietveld methods because it does not form large, well-ordered crystals. A schematic structure of sepiolite, $Mg_4Si_6O_{15}(OH)_2 \cdot 6H_2O$, was proposed by Nagy and Bradley (1955) and Brauner and Preisinger (1956) using so-called fiber diagrams (two-dimensional data). However, the details of the atom distribution in the structural tunnels between the silicate chains remained poorly understood. In what constituted essentially the first *three-dimensional* structural study of sepiolite, Bish and Post (1987) studied a sample from Durango, Mexico, using a combination of Rietveld and difference-Fourier methods with X-ray powder diffraction data (Figure 10). Using the basic framework structure determined by Brauner and Preisinger (1956), they located and refined the positions of all water molecules postulated by Brauner and Preisinger, with the addition of two other water molecules. The "water of crystallization" site is fully occupied, whereas the remaining water sites (as judged by interatomic distances) are only partially occupied. Their results do not agree in detail with the electron diffraction results of Rautureau and Tchoubar (1974), and several of the split-atom positions proposed by Rautureau and Tchoubar were not confirmed. The greatest difficulty faced in interpreting the ΔF maps was deciding which peaks were real; all of the tunnel occupants are low atomic-number atoms, and most or all of the sites are probably partially occupied. In cases such as this, one must rely on crystal-chemical reasoning, using information such as bond distances and observed stoichiometry. Like the results of Adams and co-workers, these studies of sepiolite are examples of the utility of the combination of Rietveld and difference-Fourier methods for partial structure solution. Even partially occupied water sites in relatively poorly ordered materials can be located.

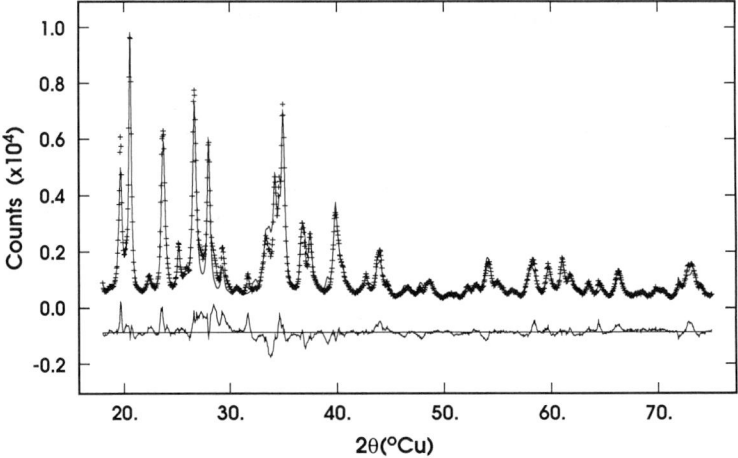

Figure 10.   Observed (+) and calculated (solid line) X-ray diffraction data (CuKα) for sepiolite. Lower curve is the difference between observed and calculated profiles.

Quantitative Analysis

Quantitative phase analysis using calculated patterns is a natural outgrowth of the Rietveld method because the refined scale factor for each crystalline phase is related to the amount of the material present in a mixture. The methodology involved in quantitative analysis using the Rietveld method is analogous to conventional single-phase Rietveld refinement, and the quantitative analysis theory is identical to that implemented in most conventional quantitative analyses (*e.g.*, Klug and Alexander, 1974; Cullity, 1978).

Theory. The integrated intensity of X-rays diffracted by a randomly oriented infinitely thick polycrystalline sample in flat-plate geometry utilizing a diffracted-beam monochromator can be written for a particular reflection as

$$I_{hkl} = \left\{ \frac{I_o A \lambda^3}{32\pi r} \left(\frac{\mu_o}{4\pi}\right)^2 \left(\frac{e^4}{m^2}\right) \left(\frac{1}{2\mu}\right) \left(\frac{1}{V^2}\right) \right. \tag{7}$$
$$\left. \times \left[|F|^2 p \frac{1+\cos^2 2\theta \cos^2 2\theta_m}{\sin^2 \theta \cos\theta} \right] e^{-2m} \right\}_{hkl}$$

where the subscript hkl denotes the dependence of particular terms on the Bragg reflection hkl. Here I = integrated intensity per unit length of diffraction line (joules sec^{-1} m^{-1}), I_0 = intensity of incident beam (joules sec^{-1} m^{-2}), A = cross-sectional area of incident beam (m^2), λ = wavelength of incident beam (m), r = radius of diffractometer circle (m), $\mu_0 = 4\pi \times 10^{-7}$ m kg C^{-2}, e = charge on electron (C), m = mass of electron (kg), V = volume of unit cell (m^3), F = structure factor, p = multiplicity factor, θ = Bragg angle, e^{-2m} = temperature factor, and μ = linear absorption coefficient (m^{-1}), which enters as the absorption factor $1/2\,\mu$ (Cullity, 1978). The term $2\theta_m$ refers to the diffraction angle of the diffracted-beam monochromator crystal. The constant (K) and variable (R_{hkl}) terms in (7) can be separated by defining

$$K = \left(\frac{I_o A \lambda^3}{32\pi r}\right)\left(\frac{\mu_o}{4\pi}\right)^2\left(\frac{e^4}{m^2}\right) \tag{8}$$

and

$$R_{hkl} = \left(\frac{1}{V^2}\right)\left[|F|^2 p\left(\frac{1+\cos^2 2\theta \cos^2 2\theta_m}{\sin^2 \theta \cos\theta}\right)e^{-2m}\right]_{hkl}. \tag{9}$$

Equation (7) can now be written in terms of (8) and (9) as

$$I_{hkl} = K(1/2\mu)R_{hkl}. \tag{10}$$

In a mixture, the intensity of the hkl reflection from the α phase is given as

$$I_{\alpha,hkl} = C_\alpha K(1/2\mu_m)R_{\alpha,hkl}, \tag{11}$$

where C_α is the volume fraction of the α phase and μ_m is the linear absorption coefficient of the mixture. In terms of weight fraction (W_α), Eq. (11) can be rewritten as

$$I_{\alpha,hkl} = \left(\frac{W_\alpha}{\rho_\alpha}\right) K\left(\frac{\rho_m}{2\mu_m}\right) R_{\alpha,hkl}, \tag{12}$$

where ρ_α is the density of phase α and ρ_m is the density of the mixture. The Rietveld scale factor, S, in Eq. (2) includes all of the constant terms in Eq. (7) and for X-rays can be written

$$S = \frac{K}{V^2 \mu}, \tag{13}$$

where V is the unit-cell volume and μ is the linear absorption coefficient for the sample. Thus, for a multi-phase mixture, Eq. (2) can be rewritten summing over the p phases in a mixture (*e.g.*, Hill and Howard, 1987) as

$$y_i(c) = y_{ib}(c) + \sum_p S_p \sum_k p_{kp} L_{kp} |F_{kp}|^2 G(\Delta\theta_{ikp}) P_{kp} . \tag{14}$$

The scale factor for each phase can now be written

$$S_\alpha = \frac{C_\alpha K}{V_\alpha^2 \mu_m}, \tag{15}$$

where C_α is the volume fraction of the α phase, V_α is the unit-cell volume of phase α, and μ_m is the linear absorption coefficient of the mixture. Recasting Eq. (15) in terms of weight fractions and the mass absorption coefficient of the mixture we obtain

$$S_\alpha = \frac{W_\alpha K}{(\rho_\alpha V_\alpha^2 \mu^*)}, \tag{16}$$

where μ^* is the mass absorption coefficient of the sample, W_α is the weight fraction of phase α, and ρ_α and V_α are the density and unit cell volume, respectively, of phase α. Alternatively, the unit-cell volume can be incorporated into the variable, phase-specific parameters as outlined by Bish and Howard (1988).

The scale factors contain the desired weight fraction information in a Rietveld analysis of a multicomponent mixture. However, the value of K and the sample mass absorption coefficient cannot easily be determined so that an analysis of an unknown sample is usually performed by constraining to unity the sum of the weight fractions of the phases considered. Thus, for a two-phase mixture,

$$W_\alpha = \frac{W_\alpha}{(W_\alpha + W_\beta)}. \tag{17}$$

Equation (17) can be solved for the weight fractions of the α and β phases to yield an expression for the weight fraction of phase α in terms of the scale factor information determined in the Rietveld analysis,

$$W_\alpha = \frac{S_\alpha \rho_\alpha V_\alpha^2}{(S_\alpha \rho_\alpha V_\alpha^2 + S_\beta \rho_\beta V_\beta^2)} \tag{18}$$

In general, the weight fraction for the ith component in a mixture of n phases can be obtained from

$$W_i = \frac{S_i \rho_i V_i^2}{\sum_j S_j \rho_j V_j^2}. \tag{19}$$

This method is exactly analogous to the adiabatic principle of Chung (1974b) in which reference intensity ratios are measured prior to analysis. The Rietveld method calculates absolute intensities to obtain intensities on an absolute scale rather than measuring reference intensity ratios.

A second method of Rietveld quantitative analysis requires that a known weight fraction of a crystalline internal standard be added to the unknown mixture. The internal standard can be any well-crystallized material that is readily available in pure form. Bish and Howard (1988) used Si, but corundum appears to be preferable (Linde A corundum) (Bish and Post, 1993). If W_α is known, then an additional parameter (C) can be evaluated from the internal standard:

$$C = \frac{S_\alpha \rho_\alpha V_\alpha^2}{W_\alpha} = \frac{K}{\mu^*} \tag{20}$$

This parameter can then be used to determine the weight fractions of other phases in the sample. For example, the weight fraction for the β phase is determined by

$$W_\beta = \frac{S_\beta \rho_\beta V_\beta^2}{C}, \tag{21}$$

where S_β is the refinable scale factor for the β phase, ρ_β is calculated from the composition and cell parameters of phase β, and C is determined using an internal standard and Eq. (20). Therefore, the weight fraction of phase β (W_β) can be easily determined. This second method does not constrain the sum of the weight fractions, as does the first method, and it is analogous to internal-standard methods commonly used to perform quantitative analysis such as the matrix-flushing method of Chung (1974a). The total weight fraction of any amorphous components (or other phases neglected in the refinement) can thus be determined with this method if the amorphous profile can be fit with the Rietveld background polynomial and if the amorphous content is significant (greater than 5-10%). The difference between the sum of the weight fractions of the crystalline components and 1.0 is the total weight fraction of the amorphous components.

O'Connor and Raven (1988) used this method in slightly modified form. They determined the constant parameter, K, from a single sample and used the refined unit-cell parameters and cell contents to evaluate density and volume. The mass absorption coefficients were computed using the known compositions of the two phases in their single sample. It appears that some pitfalls may exist with this approach in light of their results on a 50:50 quartz:corundum mixture in which they concluded that their quartz contained 18% amorphous component.

Bish

Application. Application of the Rietveld quantitative analysis method has been demonstrated on a variety of geological systems by Bish and Post (1993), but it is worth describing the application to mixtures containing clay minerals because significantly different problems arise. As emphasized above, the Rietveld method cannot be applied rigorously to materials exhibiting non-Bragg diffraction effects, such as many clays and layer silicates. With this *caveat*, the method can be successfully applied to a variety of important systems containing clays and clay minerals. Any materials in a mixture exhibiting non-Bragg diffraction effects can be analyzed by difference, using an internal standard. Alternatively, such materials can be included in a refinement in unusual cases. Guthrie and Bish (unpublished) successfully analyzed a synthetic mixture of dehydrated halloysite and corundum, assuming the dickite structure for halloysite. Only the scale factor and lattice parameters for halloysite were allowed to vary, and all other phase-specific parameters for halloysite were kept constant. The quantitative analysis was successful apparently because the *useful* observations for halloysite consisted only of 00l reflections and the sample exhibited no preferred orientation of crystallites. The fit between observed *hk* bands and calculated *hkl* reflections was so poor that the bands did not appreciably affect the scale factor, *i.e.*, there were significant errors above and below the calculated diffraction pattern. However, the poor fit between observed and calculated reflections produced a very large agreement factor. Such a situation may not always exist for samples containing clay minerals, so these mixtures should always be analyzed with caution.

The analyses of bauxite samples presented by Bish and Jones (1991) illustrate a successful application of the method to poorly crystalline soil samples. They analyzed a suite of Hawaiian bauxites taken from the surface to 290-cm depth using CoKα radiation. Although these samples represent complex (up to eight phases) and poorly crystalline mixtures, the refinements converged to consistent values and the resultant fits were excellent (Figure 11). The samples contained no materials exhibiting any two-dimensional diffraction effects so the Rietveld method could be applied without significant assumptions. The quantitative results were consistent with chemical data, and integrated peak intensities for the goethite 110 reflection found by profile fitting correlated very well with the quantitative results from Rietveld refinement ($r^2 = 0.96$). Quantitative analyses for 12 samples are shown in Table 1. In addition, the Rietveld refinements yielded precise unit-cell parameters for the major phases; for example, the goethite parameters suggested an Al content between 28 and 35%, and the magnetite/maghemite parameters showed that they were 75-80% oxidized towards end-member maghemite (Table 2). The plots of X-ray diffraction data clearly showed that the goethite reflections were anisotropically broadened, so refinements were conducted using the program GSAS (Larson and Von Dreele, 1988) that can explicitly model anisotropic crystallite-size and strain broadening. The results of the GSAS refinements suggested that the goethite reflections had a significant strain-broadening component in addition to a small crystallite size component. The crystallite-size broadening was primarily perpendicular to the (001) plane, whereas the strain broadening was principally in the *X-Y* plane. The major phases in these samples did not exhibit significant preferred orientation, probably because of their fine-grained nature.

Interestingly, the Rietveld method can also be easily adapted to the analysis of trace phases. When the full pattern in Figure 12 was originally used in a multi-phase Rietveld quantitative analysis, refined amounts of erionite (0.14%) were consistently above the known added amount of erionite (0.05%). Further analysis of the "pure" analcime with the Rietveld method revealed the presence of ~0.07% erionite and ~0.43% illite, missed in the original analyses. The final Rietveld quantitative analyses are thus consistent with the amount of erionite in the analcime-erionite standard.

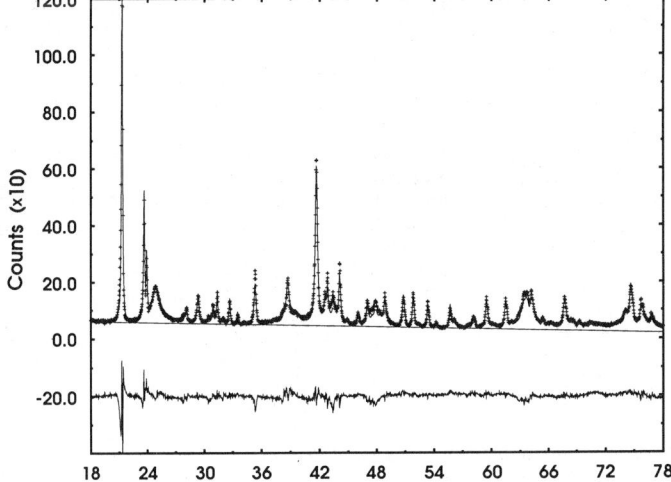

Figure 11. Observed (+) and calculated (solid line) X-ray diffraction data (CoKα) for Hawaiian bauxite sample from 260- to 290-cm depth. Lower curve is the difference between observed and calculated profiles.

Table 1. Quantitative analyses (wt.%) of Hawaiian bauxites as a function of depth.

Mineralogy	Hematite	Goethite	Gibbsite	Magnetite	Anatase	Rutile	Ilmenite
Sample Depth (cm.)							
0-10[1]	4.1	38.6	39.6	6.5	6.2	0.2	0.3
10-20	2.5	43.3	40.0	7.5	6.4	0.1	0.2
20-30	3.2	45.6	37.7	6.5	6.7	0.2	0.2
30-50	6.2	32.8	46.8	8.4	5.2	0.2	0.3
50-70	4.0	30.7	53.3	6.4	5.2	0.2	0.2
70-90	6.0	31.8	48.4	8.5	5.0	0.2	0.2
90-100	3.5	24.8	53.1	14.7	3.9	0.03	0.03
110-120	9.5	18.9	47.2	20.4	3.7	0.2	0.02
130-160	3.5	40.7	42.1	8.4	4.9	0.1	0.2
170-200	6.6	26.3	49.6	14.2	2.8	0.2	0.4
220-260	5.0	19.7	60.0	11.5	3.5	0.1	0.2
260-290	7.2	23.7	51.9	13.3	3.2	0.4	0.4

[1]The 0-10 cm sample also contains 4.5% quartz which was not detected in all other samples

Table 2. Unit-cell parameters (Å) for goethite and magnetite (maghemite) for Hawaiian bauxite samples.

	Goethite			Magnetite
	a	b	c	a
Sample Depth (cm.)				
0-10	4.582(1)[1]	9.827(3)	2.9731(6)	8.3561(5)
10-20	4.584(1)	9.843(3)	2.9750(7)	8.3588(8)
20-30	4.583(1)	9.843(3)	2.9758(7)	
30-50	4.587(2)	9.849(5)	2.973(1)	
50-70	4.584(2)	9.855(5)	2.972(1)	
70-90	4.587(2)	9.864(5)	2.976(1)	
90-100	4.580(2)	9.812(7)	2.969(2)	
110-120	4.595(4)	9.84(1)	2.964(2)	
130-160	4.581(2)	9.841(5)	2.9699(9)	
170-200	4.589(3)	9.87(1)	2.966(2)	8.3631(6)
220-260	4.583(4)	9.92(1)	2.969(2)	
260-290	4.591(3)	9.91(1)	2.964(2)	

[1] Numbers in parentheses represent estimated standard deviations in the last quoted decimal place.

Refinement of Unit-Cell Parameters

The Rietveld method minimizes the differences between observed and calculated diffraction patterns at every data point rather than relying on simpler methods that use only peak positions. In addition, calculated peak positions are constrained by the model symmetry and by a set of unique unit-cell parameters, resulting in refined unit-cell parameters that are most consistent with the observed data. The calculated diffraction pattern also incorporates all possible reflections contributing to a peak that might shift its position slightly but may be too weak or too closely overlapped to be resolved. Therefore, the method yields unusually precise information on unit-cell parameters and reflection positions. Figure 13 illustrates a region of the diffraction pattern of clinoptilolite, showing the Rietveld fit to a composite three-peak region. The refinement results shown in the figure show that this three-peak region is instead composed of at least six reflections. Obviously, this overlap (common in powder diffraction patterns) would make unique indexing of the peaks in the pattern difficult and would preclude accurate unit-cell parameter refinement by conventional methods.

The full-pattern fitting routine used in the Rietveld method also provides the opportunity to correct analytically for a number of the most serious systematic errors affecting peak positions, including specimen displacement, specimen transparency, and zero-2θ offset. These effects, and corrections for them, are discussed by Klug and Alexander (1974) and Jenkins (1989). Incorporation of some or all of these effects allows the Rietveld method to provide unit-cell

Studies of Clays Using the Reitveld Method

parameters not only of high precision but also of high accuracy. The Rietveld method appears to be superior to any other conventional method for determining unit-cell parameters.

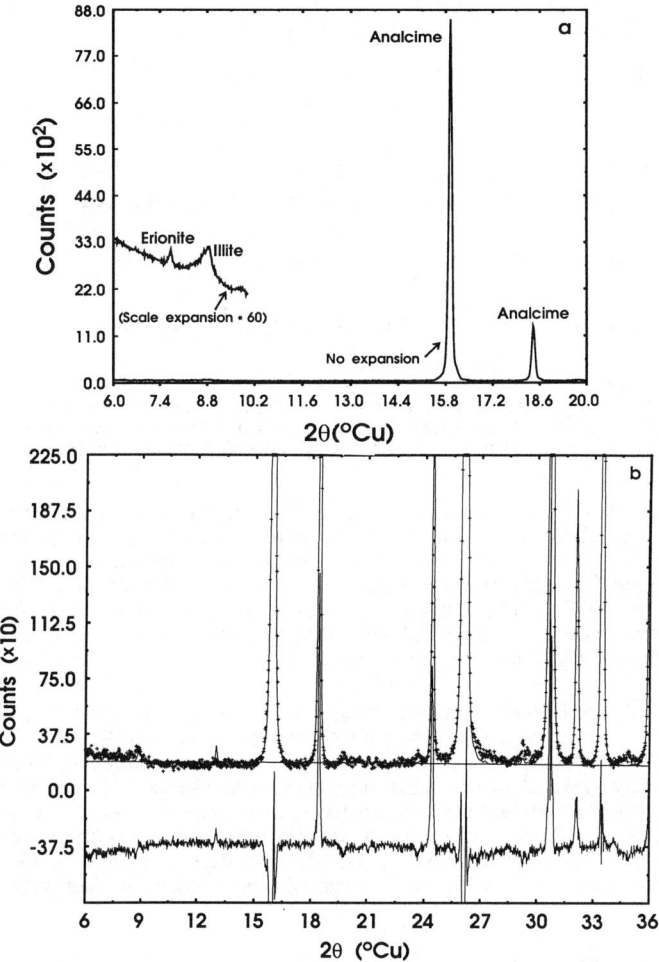

Figure 12. a) Observed X-ray powder diffraction pattern of a synthetic analcime-erionite mixture, showing the presence of ~0.14% erionite and ~0.43% illite; and b) observed (+) and calculated (solid line) X-ray diffraction data (CuKα) for the analcime-erionite mixture. Lower curve is the difference between observed and calculated profiles.

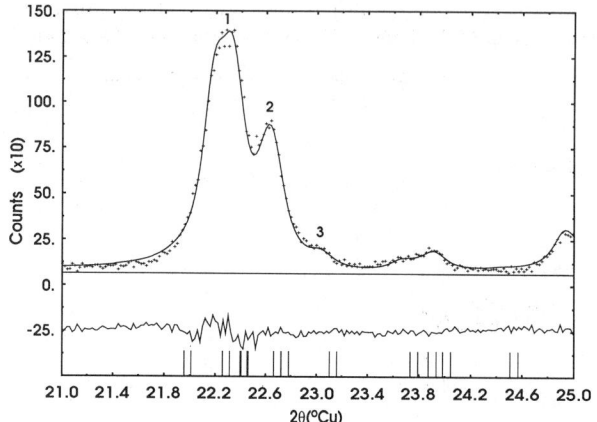

Figure 13. A portion of the observed (+) and calculated (solid line) powder X-ray diffraction pattern of clinoptilolite, showing the fit to a three-peak (peaks are labeled) region. The lower curve shows the difference between observed and calculated patterns, and vertical marks (in pairs) at the bottom indicate the positions of allowed $K\alpha 1$ and $K\alpha 2$ reflections (R_{wp} = 14.7%, R_{exp} = 7.5%).

Post and Bish (1989) illustrated that consistent unit-cell parameters could be obtained for Si, even for samples displaced from the focal plane by up to 8 mils. They also gave refined unit-cell parameters for corundum, hematite, and quartz. Use of a specimen transparency correction in addition to a specimen displacement correction can yield even greater accuracy. Refinement of the unit-cell parameter of NIST 640a Si without either correction gave a value of a = 5.43073(2)Å and use of both corrections gave a value of 5.43087(3)Å, comparing well with the certified value of 5.430825(36)Å (statistically identical at the 1σ level).

Application of the Rietveld method to synthetic Mn-goethite samples of varying Mn substitution (Bish and Ebinger, 1989) gave unit-cell values of a = 4.6056(4)Å, b = 9.9657(8)Å, and c = 3.0210(2)Å, significantly more precise than previous measurements. Precise values were obtained in spite of the presence of very broad reflections and a hematite impurity. It is difficult to assess the accuracy of these values because the true values are not known. The goethite a and c dimensions decreased by 0.039Å and 0.096Å, respectively, and the b dimension increased by 0.474Å as the Mn mole fraction increased to 0.47. These changes in unit-cell parameters appear to be consistent with increasing Jahn-Teller distortion in the octahedral sites with increasing Mn substitution.

Analysis of Peak Broadening

Because the Rietveld method explicitly fits all observed diffraction peaks, considerable information on the profiles can be extracted, particularly if the capability exists to fit anisotropically broadened reflections. This treatment can potentially go far beyond the familiar Scherrer method which relates peak broadening to crystallite size. An analytical peak-shape function is explicitly included in all Rietveld refinements, a function that as accurately as possible represents the observed peak shapes. Although earlier Rietveld refinement programs used simple profile functions, such as a Gaussian or Cauchy profile, more recent programs use more complex functions such as a pseudo-Voigt (combination of Gaussian and Cauchy) or Pearson *VII* function (see Young and Wiles, 1982, for a description of many profile functions). These profile functions have commonly been combined with an angle-dependent peak-width function, the most common being the formulation of Caglioti *et al.* (1958) relating peak width, H, to angle:

$$H = U\tan^2\theta + V\tan\theta + W, \qquad (22)$$

where H is the peak full width at half-maximum height, and U, V, and W are refinable parameters.

Instead of the Caglioti *et al.* (1958) formulation, the program GSAS (Larson and Von Dreele, 1988) uses a multi-term Simpson's-rule integration of the pseudo-Voigt function that contains separate expressions for Gaussian and Lorentzian broadening and incorporates the ability to accommodate anisotropic broadening. The Lorentzian component is parameterized in terms of $1/\cos\theta$ and $\tan\theta$. As shown below, this formulation allows separation of crystallite size and strain contributions to reflection broadening.

Crystallite-size broadening causes all reflections in reciprocal space to be broadened alike, i.e., $\Delta d^* =$ constant. Because $d^* = d^{-1}$, $\Delta d^* = -d^{-2}\Delta d = \Delta d/d^2 =$ constant, crystallite-size broadening is related to 2θ broadening by:

$$\Delta d/d^2 = \frac{\Delta\theta \cot\theta}{d} = \text{constant}. \qquad (23)$$

Using Bragg's law, and the equality $2\Delta\theta = \Delta 2\theta$,

$$\Delta d/d^2 = \frac{\Delta 2\theta \cot\theta \sin\theta}{\lambda}. \qquad (24)$$

Broadening is then:

$$\Delta 2\theta = \frac{\lambda(\Delta d/d^2)}{2\cos\theta} \qquad (25)$$

Thus, crystallite size information can be extracted from a refinement if profile coefficients are parameterized to relate peak breadth to $1/\cos\theta$.

Bish

Strain (essentially a distribution of unit-cell parameters) broadening causes a shift in diffraction line position, and broadening in reciprocal space is a function of d^*, i.e, $\Delta d^*/d^*$ = constant. From the above relationships we can show therefore that $\Delta d/d$ = constant. Thus strain broadening in real space (as opposed to reciprocal space) is related to 2θ broadening by:

$$\Delta d / d = \Delta 2\theta \cot\theta = \text{constant} \tag{26}$$

or

$$\Delta 2\theta = (\Delta d/d)\tan\theta \ (\Delta 2\theta \text{ in radians}). \tag{27}$$

Thus, if the profile coefficients are parameterized to relate peak breadth to tanθ, information on lattice strain can be extracted from a refinement. Broadening of a reflection (in Δ2θ) due to small crystallite size varies as the inverse of cosθ, whereas strain broadening (again in Δ2θ) varies as tanθ. Thus, determination of the pseudo-Voigt breadth terms related to 1/cosθ and tanθ can provide information on crystallite-size and strain broadening, respectively.

Goethite. The Rietveld refinements of goethites (FeOOH) by Bish and Ebinger (1989) and Ebinger and Bish (1990) are good examples of the extraction of information on the sources of peak broadening in a mineral. Bish and Ebinger (1989) used these methods with X-ray powder diffraction data in a study of goethite and Mn-goethites to determine not only structural information but insights into the causes of peak broadening for the samples (Figure 14). Their refinements illustrated that increasing Mn substitution in goethite causes a decrease in the *a* and *c* unit-cell parameters and an increase in the *b* parameter. Atomic positions obtained for goethite were significantly different from limited literature data, and they showed that the environment of the metal site progressively changes as the amount of Mn substitution increases. The (Fe,Mn)-O,OH distances systematically changed with an increase in Mn mole fraction. The two apical metal-O,OH distances lengthened significantly whereas the equatorial metal-O,OH bonds shortened with an increase in Mn mole fraction, consistent with a Jahn-Teller distortion that increased with incorporation of Mn. The relative changes in the unit-cell parameters are readily explained by these atomic changes. A new and interesting result of their study is the indication from the refined profile parameters that the principle cause of broad reflections in the X-ray diffraction patterns is lattice strain rather than small crystallite size. These parameters implied that crystallite size either remained constant or increased with increasing Mn mole fraction. However, lattice strain increased with increasing Mn, reaching values in excess of 5% strain at 0.47 Mn. Results also indicated that strain is greatest in the *X-Y* plane. Bish and Ebinger concluded that the highly strained lattice is probably the reason for the observed structural break between goethite and groutite (MnOOH). These results obtained from the observed profiles are in direct opposition to published results on goethite; all previous investigators had assumed the cause of broad reflections to be due to small crystallite sizes and had applied the Scherrer equation. Because of the extreme breadth of the goethite reflections and the presence of overlapping reflections (including contributions from hematite in the low-Mn materials) in the diffraction pattern, extraction of this information using other methods such as profile fitting would have been virtually impossible.

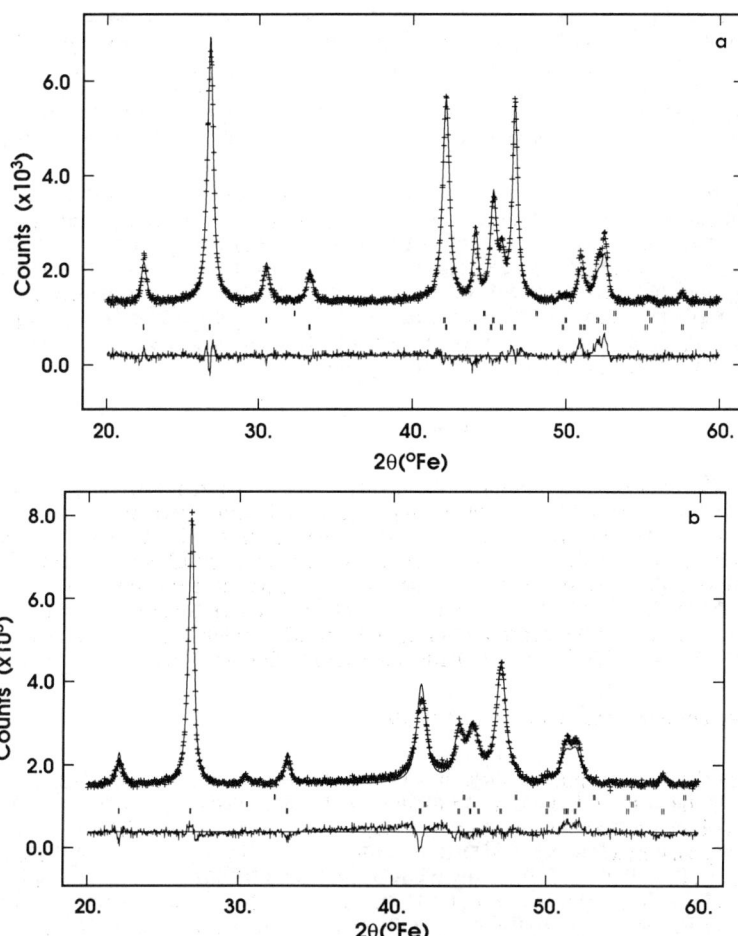

Figure 14. a) Observed (+) and calculated (solid line) X-ray powder diffraction data (FeKα radiation) for synthetic goethite, FeOOH. Lower curve is the difference between observed and calculated profiles, and the tic marks indicate positions for allowed FeKα1 and Kα2 reflections of hematite (upper) and goethite (lower). Goethite reflections are primarily crystallite-size broadened. b) Observed (+) and calculated (solid line) X-ray powder diffraction data (FeKα radiation) for synthetic Mn-goethite, (Fe,Mn)OOH. Lower curve is the difference between observed and calculated profiles, and the tic marks indicate positions for allowed FeKα$_1$ and Kα$_2$ reflections of hematite (upper) and goethite (lower). Goethite reflections have a significant strain-broadening component.

The Rietveld refinements of Ebinger and Bish (1990) on Cr-goethites are similar to those of Bish and Ebinger (1989) on Mn-goethites. The unit-cell parameters for Cr-goethites decreased only a minor amount as Cr substitution increased, due to the similarity in size between Cr^{+3} and Fe^{+3}. In contrast with the results for Mn-goethites, the broadening of reflections in diffraction patterns for Cr-goethites appeared to result primarily from small crystallite sizes rather than strain. This is consistent with the similarity in ionic size and the lack of Jahn-Teller distortions for Cr^{+3}.

Clearly there are many potential applications of the Rietveld method to clay minerals in extracting information on the sources of peak broadening, because many clay minerals exhibit significantly broadened reflections. For example, the refinement of kaolinite by Bish and Von Dreele (1989) suggested that the kaolinite reflections were broadened by both crystallite size and strain effects, both of which were anisotropic. The strain broadening was greater along [001] than within the *X-Y* plane, consistent with a finite distribution in the *c* unit-cell parameter. Based on these results and those for goethite, it is probably fallacious to neglect the effects of strain broadening in studies of clay mineral diffraction peaks.

SAMPLE REFINEMENT

A typical Rietveld refinement is a complex undertaking, involving numerous variable parameters related to background, profile shapes, unit-cell parameters, preferred orientation, and structure. Due to the often large number of variable parameters, it is very easy to "get a pie in the face" by allowing parameters to vary too early in the refinement or allowing unjustified parameters to vary. In this respect, Rietveld refinements are very different from conventional single-crystal refinements in which individual reflection intensities or structure factors are used as observations. Because of the difficulties encountered during a Rietveld refinement, it is worthwhile to outline a typical and recommended refinement scheme that should minimize difficulties.

Sample Preparation and Data Collection

Perhaps the most important step in a Rietveld refinement is sample preparation and data collection as is the case with many instrumental methods of analysis. It helps to remember that a typical data collection may take as long as a day or two, unattended, whereas a typical Rietveld refinement may take much longer and require frequent user interventions. Thus, time spent on data collection is time well spent. It is always advantageous to use as pure a sample as possible, although most modern Rietveld programs incorporate the ability to treat multiple phases. Otherwise, the foremost requirement for a Rietveld refinement is *accurate* intensity data, so attempts must be made to obtain adequate particle statistics (using a sufficiently fine particle size) and to minimize preferred orientation. Bish and Reynolds (1989) suggested that particle sizes be at least as small as 10 μm for materials of average linear absorption coefficient, and smaller if high linear absorption coefficient materials are used. As outlined by Bish and Reynolds, a variety of sample-mounting methods exist to reduce preferred orientation, but spray drying appears to produce the most "random" sample mount. If the Rietveld method is being used only for unit-cell parameter determination from a complex pattern, these requirements can be relaxed; relatively poor quality data can be used as long as no uncorrected systematic errors are present that can influence the reflection positions.

For both X-ray and neutron studies, it is recommended that data be collected out to the high-angle limit of the instrument. This will provide much more stable refinement of angle-dependent variables such as temperature factors, site occupancies, and profile parameters (yielding information on crystallite size and strain broadening). Although it is a common perception that little or no useful information is available from the high-angle portion of diffraction patterns, this is usually not true. If a sample is sufficiently ordered and exhibits only Bragg diffraction effects, high-angle data will probably be useful. The high-angle data may appear broad and poorly resolved, but this is usually due to low intensities and overlapping of numerous reflections. The low-angle cutoff in data collection should be determined based on beam coverage of the sample and on instrumental effects. Low-angle reflections can have considerable asymmetry in both X-ray and fixed-wavelength neutron data, due in part to axial divergence of the beam (incident-beam Soller slits should be used). In addition, it is often worthwhile to avoid using data in the angular range in which Lorentz and polarization effects are rapidly varying. In any X-ray experiment, the sample must be fully within the incident beam at the lowest angle of interest, although a correction can be applied to the data if the X-ray beam and sample dimensions are known. Choosing a low-angle cutoff as high as 20 °2θ seldom results in the loss of many reflections. Finally, it is important to note that a well-aligned diffractometer and well-known calibration standards are important in any Rietveld refinement. Although most Rietveld programs provide the ability to refine both zero-point and sample-displacement corrections, these two corrections are highly correlated and simultaneous refinement of both values may yield unstable results. Use of a diffractometer with no zero-point error and refinement of only a sample displacement correction is preferable.

It is possible and advantageous to obtain digital X-ray data in a variety of separate angular ranges in an attempt to acquire data with approximately constant precision. Consider a diffraction pattern in which reflections between 10 and 50 °2θ are on the average five times as intense as those between 50 and 100 °2θ, which are in turn five times stronger on average than those between 100 and 160 °2θ. An appropriate data-collection strategy would then involve collection of three separate ranges, for example counting for 2 s/step in the first range, 10 s/step in the second range, and 50 s/step in the third range. The author has obtained good results using this data-collection strategy in conjunction with a Rietveld program that facilitates the use of multiple data sets (*e.g.*, GSAS, Larson and Von Dreele, 1988). The use of multiple ranges is particularly beneficial for X-ray diffraction studies because of the decrease in atomic scattering factors with angle. However, neutron scattering lengths do not decrease as a function of angle, so the multiple-range method is not necessary in neutron diffraction studies. The author has found that the use of this data-collection method with X-ray diffraction data yields improved values for temperature factors and site occupancies, and individual temperature factors can be stably refined in favorable cases with complex structures. Post and Bish (1989) outlined the various arguments concerning choice of step sizes and count times for use in Rietveld data collection, and the interested reader should consult that discussion and the references cited therein.

Refinement Strategies

To reiterate, the approximate crystal structures of all phases in a given sample must be known before performing a Rietveld refinement. The structures, together with the unit-cell parameters and space group symmetry of each phase, make up the sample-related parameters to be used as input to any Rietveld refinement program. In addition, before beginning a refinement, the types of background and profile functions to be used must be selected; the available options will be determined by the particular Rietveld program used. Most refinements with X-ray diffraction observations today use either pseudo-Voigt or Pearson *VII* profile functions, or a more complex variant of one of these. Appropriate starting values for the profile function are what are known as instrument parameters, *i.e.*, values unaffected by any sample-related broadening. These values should be determined for an individual instrument with an appropriate material that contributes little or no sample-related broadening (*e.g.*, NIST SRM 660, LaB_6) so that instrument-related profile broadening can be evaluated and incorporated into the analysis of profiles. Table 3 illustrates a typical input file for a one-phase refinement of the structure of goethite for the DBW program of Wiles and Young (1981). This table is not meant to provide a template from which to prepare a Rietveld input file, but it is presented to illustrate the type of information typically required to perform a Rietveld refinement. As is obvious, there are numerous control parameters that define various instrumental, data, and refinement-related input values. Input for each phase consists of a title line, a line listing the number of atoms in the asymmetric unit, and, possibly, a preferred orientation direction. Following this information is a line containing the space group, and all atoms in the asymmetric unit are listed next. Input for each atom consists of one line providing the x, y, and z atomic parameters, an isotropic temperature factor, and the site occupancy times the site multiplicity. The second line for each atom consists of six anisotropic temperature factor values, often set to zero in a Rietveld refinement. The next line contains the scale factor and an overall temperature factor, followed on the next line by the U, V, and W parameters of Caglioti et al. (1958), a profile number (in this case Pearson *VII*), the value of the Pearson *VII* coefficient and its angle-dependence. Starting unit-cell parameters are contained on the next line, followed by a line containing information on a preferred-orientation correction and an asymmetry correction. The remainder of the file contains a variety of what are termed codewords that define which refinable parameters will be refined. Many varieties of input files exist for Rietveld refinement programs, and this table simply provides an example of the type of structural information and refinement parameters that are required.

Although the refinement strategy outlined here works well for the author, other similar strategies are used, and this procedure is by no means universally accepted in detail. After preparation of an input file containing the sample-related parameters and appropriate background and profile parameters, refinement can be commenced by allowing individual scale factors for each phase to vary. This stage of refinement is often combined with variation of background parameters (if fixed background points are not used). The aim at this point is to bring observed and calculated diffraction patterns to the same scale and to generate a background that closely matches the observed background. If a refined scale factor is too low, calculated intensities will be uniformly low by a *constant factor*; non-uniform intensity discrepancies can result from a variety of factors, including incorrect temperature factors, incorrect Lorentz-polarization factor, and errors in site occupancies or atomic positions. If the starting structural and unit-cell parameters are reasonably close to the true values, this first stage of refinement is usually successful. On some occasions,

Studies of Clays Using the Reitveld Method

various parameters may require damping to remain stable, *i.e.*, applying less than the full calculated shift to the parameter during each least-squares cycle. Throughout every stage of a refinement, it is important that progress be monitored by examining plots of observed and calculated data, particularly if any difficulties are encountered. Thus, it is imperative that quick plotting capabilities be available, *e.g.*, on a graphics terminal. Any problems with the refinement, *e.g.*, improper background parameters or scale factor, incorrect unit-cell parameters, incorrect profile parameters, or a missing phase, are usually readily apparent in the plots. It should also be obvious that a Rietveld refinement usually progresses in discrete steps, with refinement of a given set of parameters to convergence, addition of another parameter, and re-refinement to convergence. Virtually any variable may refine to unreasonable values (*e.g.*, negative or very large temperature factors, negative profile parameters), and care should be exercised when adding additional refinable parameters.

Table 3. Sample goethite input file for the DBW Rietveld program.

(title)	Goethite
	0 6 1 0 0 0 0 0 1 0
(control params)	001011000100
	1.5406 1.5444 0.5000 0.0000 15.0000 0.7992 1.0000 45.0000
	120.400.700.700.700.70 15.0000 0.0200 91.9000
	19 0 0.0000 0 0.0000 0 0.0000 0 0.0000
(displacement)	0.000000 0.00 0.02885 111.00
(background coeff)	87.8132 3.00396 0.823991E-010.000000E+000.000000E+000.000000E+00
(bkg. codewords)	21.00 31.00 41.00 0.00 0.00 0.00
(phase name)	Goethite
(# atoms)	3 0 00.000.000.00
(space group)	P B N M
	FE FE+2 0.04607 0.85100 0.25000 0.80000 0.50000
(atoms, temperature	0.00000 0.00000 0.00000 0.00000 0.00000 0.00000
factors, &	O O-1 0.79557 0.18709 0.25000 1.20000 0.50000
occupancies)	0.00000 0.00000 0.00000 0.00000 0.00000 0.00000
	OH O-1 0.15060 0.07292 0.25000 1.20000 0.50000
	0.00000 0.00000 0.00000 0.00000 0.00000 0.00000
(scale factor)	188.342 1.08108
(profile parameters)	0.00000 0.51488 0.07053 6 1.2088 0.0000
(unit-cell	4.60989 9.95492 3.01957 90.0000 90.0000 90.0000
parameters)	0.0000 0.0000 0.0135
	141.00 151.00 0.00 201.00 0.00
	0.00 0.00 0.00 0.00 0.00 0.00
(codewords)	161.00 171.00 0.00 211.00 0.00
	0.00 0.00 0.00 0.00 0.00 0.00
	181.00 191.00 0.00 221.00 0.00
	0.00 0.00 0.00 0.00 0.00 0.00
	11.00 131.00
	0.00 91.00 81.00 101.00 0.00
	51.00 61.00 71.00 0.00 0.00 0.00
	0.00 0.00 121.00

Once all profile-related parameters, unit-cell parameters, and atom positions have been stably refined, it should be possible to refine variably occupied sites if the scattering power of the occupants differ, *e.g.*, Fe and Mg for X-rays. Site-occupancy refinement of atoms of similar scattering power, *e.g.*, Al and Si with X-rays, should not usually be attempted because unpredictable results will probably occur. If data are available over a wide angular range, it is usually possible to refine individual temperature factors, particularly if like atoms are constrained to have the same shifts. However, simultaneous refinement of temperature factors and site occupancies often yields unstable results due to correlations between the two. If partial structure solution is desired, *e.g.*, determination of the location of an exchangeable cation in a zeolite, the recommendations in Post and Bish (1989) should be followed to avoid biasing the refinement.

The Rietveld method is more prone to yielding inaccurate refined structures than are single-crystal methods, and it is always important to assess the reliability of a final structure. The weighted profile residual [Eq. (1)] provides a means of assessing the quality of the fit, but it does not allow assessment of the statistical significance of the refined structure. A variety of statistical parameters, including the Durbin-Watson *d* statistic and the reduced χ^2 statistic (see Post and Bish, 1989) have been proposed that allow one to obtain an idea of how significant a final result is. Irrespective of what R_{wp} value is obtained, it is important to assess how crystal-chemically reasonable a final structure is. Generally, this involves a comparison of the observed bond distances and angles with published results from comparable structures. If observed bond distances deviate significantly from accepted values, the refined results should be suspect. In rare (?) cases, the refined structure may represent a false-minimum result, although poor and inaccurate structures usually result from various combinations of incorrect starting model, poor-quality data, incorrect refinement strategies, and uncorrected systematic errors.

When presenting results, it is important to provide all information pertinent to the refinement, including the type of profile function used, the value and type of preferred orientation correction, and the type of temperature factors (individual or overall) used. Observed, calculated, and difference curves should also be provided in plots, and it is essential to state whether background has been subtracted or not when using neutron powder diffraction data. Markers showing the calculated positions of possible reflections are often useful but are less so for complex patterns. The wavelength, scan range, count time per step, and temperature used in data collection should always be provided. Finally, in order to allow the reader to assess the quality of the refinement, both R_{wp} and expected R (R_{exp}) value should be provided (in addition to one of the useful statistical parameters, *d* or reduced χ^2). R_{exp} is essentially the minimum R_{wp} that can be obtained and is one measure of how much information remains unresolved in the data after refinement.

CONCLUSIONS

Rietveld refinement is clearly a versatile and powerful method that can be applied to solve a variety of problems in clay mineralogy if the user is aware of the appropriate sample limitations and the assumptions inherent to the method. The method can be applied to well-ordered clays and many other fine-grained minerals, such as many oxides and hydroxides, without significant assumptions and it can provide information that was previously beyond reach. This information includes accurate and precise unit-cell parameters, separation of crystallite-size and strain broadening effects, quantitative multicomponent analysis, partial structure solution, and full structure refinement. The method is particularly important for such materials because it allows analysis of very finely crystalline minerals that may be unavailable in single-crystal form.

The Rietveld method is limited in its application to disordered clays and clay minerals because it implicitly assumes the presence of Bragg diffraction effects only. Furthermore, it is very important to understand that two-dimensional diffraction effects cannot be approximated by broadening reflections using the profile shape functions typically available with the Rietveld method. Thus, a full three-dimensional refinement of such materials cannot be performed, but abbreviated analysis is still possible, including one-dimensional structure refinement which provides a projection of all atoms onto c. Such an analysis can provide accurate and precise information on layer repeat and can potentially provide information on site occupancies. Semi-random stacking sequences present an intermediate case, but their analysis will probably lead to a refined structure that is at least partially an artifact due to the presence of some two-dimensional diffraction effects.

ACKNOWLEDGMENTS

I am grateful to G. Guthrie for insightful comments on the manuscript and for assistance with figures. The manuscript has also benefitted significantly from comments by R. Reynolds and J. Walker.

REFERENCES CITED

Adams, J. M. (1983) Hydrogen atom positions in kaolinite by neutron profile refinement: *Clays & Clay Minerals* **31**, 352-356.

Adams, J. M. and Hewat, A. W. (1981) Hydrogen atom positions in dickite: *Clays & Clay Minerals* **29**, 316-319.

Banfield, J. F., Veblen, D. R., and Smith, D. J. (1991) The identification of naturally occurring $TiO_2(B)$ by structure determination using high-resolution electron microscopy, image simulations, and distance-least-squares refinement: *American Mineralogist* **76**, 343-353.

Bish, D. L. (1989) Rietveld refinement of the kaolinite structure at 88, 294, and 573K: Clay Minerals Society 26th Annual Meeting, Abstracts, p. 17.

Bish, D. L. (1992) Structure building with Rietveld analysis: in *Accuracy in Powder Diffraction II*, Proceedings of the International Conference May 26-29, 1992, NIST Special Publication 846, 154-164.

Bish

Bish, D. L. (1993) Applications of Rietveld methods to clays and clay minerals: in CMS Workshop Lectures, *Neutron Scattering Methods*, F. Ross and J. Walker, eds., The Clay Minerals Society, in press.

Bish, D. L. (1994) Rietveld refinement of the kaolinite structure at 1.5 K: *Clays & Clay Minerals*, in press.

Bish, D. L. and Ebinger, M. H. (1989) Rietveld refinement of synthetic goethite and Mn-substituted goethite: in *Proceedings of the 26th Annual Clay Minerals Society Meeting*, 18.

Bish, D. L. and Howard, S. A. (1988) Quantitative phase analysis using the Rietveld method: *Journal of Applied Crystallography* **21**, 86-91.

Bish, D. L. and Johnston, C. T. (1993) Rietveld refinement and Fourier-transform infrared spectroscopic study of the dickite structure at low temperature: *Clays & Clay Minerals* **41**, in press.

Bish, D. L. and Jones, R. C. (1991) Quantitative X-ray diffraction analysis of soils: Comparison of conventional curve-fitting and Rietveld full-pattern methods: in *Proceedings of the 28th Annual Clay Minerals Society Meeting*, 16.

Bish, D. L. and Post, J. E. (1987) Refinement of the sepiolite structure and location of tunnel water molecules using the Rietveld method: in *Proceedings of the 24th Annual Clay Minerals Society Meeting*, 33.

Bish, D. L. and Post, J. E. (1993) Quantitative mineralogical analysis using the Rietveld full-pattern fitting method: *American Mineralogist* **78**, in press.

Bish, D. L. and Reynolds, R. C., Jr. (1989) Sample preparation for X-ray diffraction: in *Modern Powder Diffraction*, D. L. Bish and J. E. Post, eds., Mineralogical Society of America, Washington, D. C., 73-99.

Bish, D. L. and Von Dreele, R. B. (1989) Rietveld refinement of non-hydrogen atomic positions in kaolinite: *Clays & Clay Minerals* **37**, 289-296.

Bish, D. L. and Von Dreele, R. B. (1990) The crystal structure of kaolinite including hydrogen atoms: in *Proceedings of the 27th Annual Clay Minerals Society Meeting*, 25.

Bish, D. L., Giese, R. F., Jr., and Costanzo, P. M. (1992) Crystal structure of the 8.4Å kaolinite-hydrate complex: in *Proceedings of the 29th Annual Clay Minerals Society Meeting*.

Brauner, K. and Preisinger, A. (1956) Struktur und Entstehung des Sepioliths: *Miner. Petrogr. Mitt.* **6**, 120-140.

Buseck, P. R. and Veblen, D. R. (1988) Mineralogy: in *High-Resolution Transmission Electron Microscopy and Associated Techniques,* P. Buseck, J. Cowley, and L. Eyring, eds., Oxford Univ. Press, New York, 308-377.

Cagliotti, G., Paoletti, A., and Ricci, F. P. (1958) Choice of collimators for a crystal spectrometer for neutron diffraction: *Nuclear Instrumentation* **3**, 223-228.

Catlow, C. R. A., Doherty, M., Price, G. D., Sanders, M. J., and Parker, S. C. (1986) Computer simulation studies in silicates: *Materials Science Forum* **7**, 163-176.

Chung, F. H. (1974a) Quantitative interpretation of X-ray diffraction patterns of mixtures. I. Matrix-flushing method for quantitative multicomponent analysis: *Journal of Applied Crystallography* **7**, 519-525.

Chung, F. H. (1974b) Quantitative interpretation of X-ray diffraction patterns of mixtures. II. Adiabatic principle of X-ray diffraction analysis of mixtures: *Journal of Applied Crystallography* **7**, 526-531.

Costanzo, P. M., Clemency, C. V., and Giese, R. F., Jr. (1980) Low-temperature synthesis of a 10Å hydrate of kaolinite using dimethylsulfoxide and ammonium fluoride: *Clays & Clay Minerals* **28**, 155-156.

Ebinger, M. H. and Bish, D. L. (1990) Rietveld refinement of goethite and Cr-goethite from X-ray powder data: in *Proceedings of the 27th Annual Clay Minerals Society Meeting,* 46.

Dollase, W. A. (1986) Correction of intensities for preferred orientation in powder diffractometry: Application of the March model: *Journal of Applied Crystallography* **19**, 267-272.

Giese, R. F., Jr. (1982) Theoretical studies of the kaolin minerals: Electrostatic calculations: *Bulletin Minéralogie* **105**, 417-424.

Giese, R. F. and Datta, P. (1973) Hydroxyl orientation in kaolinite, dickite, and nacrite: *American Mineralogist* **58**, 471-479.

Guthrie, G. D. and Bish, D. L. (1991) Ionic modeling of the hydrogen sites in the kaolin polymorphs: in *Proceedings of the 28th Annual Meeting of the Clay Minerals Society,* Houston, Texas, p. 63 (abstract).

Guthrie, G. D. and Bish, D. L. (1993) Ionic modeling of the hydrogen sites in the kaolin minerals: *Clays & Clay Minerals* submitted.

Guthrie, G. D. and Veblen, D. R. (1990) Interpreting one-dimensional high-resolution transmission electron micrographs of sheet silicates by computer simulation: *American Mineralogist* **75**, 276-288.

Joswig, W. and Drits, V. A. (1986) The orientation of the hydroxyl groups in dickite by X-ray diffraction: *Neues Jahrbuch fur Mineralogie Monhatshefte* 19-22.

Larson, A. C. and Von Dreele, R. B. (1988) GSAS. Generalized structure analysis system: *Los Alamos National Laboratory Report* **LAUR 86-748**, 150 pp.

Meier, W. M. and Villiger, H. (1969) Die Methode der Abstandsvergeinerung zur Bestimmung der Atomkoordinaten idealisierter Geruststrukturen: *Zeitschrift fur Kristallographie* **129**, 411-423.

Nagy, B. and Bradley, W. F. (1955) Structure of sepiolite: *American Mineralogist* **40**, 885-892.

O'Keefe, M. A. (1984) Electron image simulation: A complementary processing technique: in *Electron Optical Systems*, J. J. Hren, F. A. Lenz, E. Munro, and P. B. Sewell, eds., SEM Inc., AMF O'Hare, Chicago, IL, 209-220.

Post, J. E. and Bish, D. L. (1989) Rietveld refinement of crystal structures using powder X-ray diffraction data: In D. L. Bish and J. E. Post, Eds., *Modern Powder Diffraction*. Mineralogical Society of America Reviews in Mineralogy, **20**, 277-308.

Post, J. E. and Veblen, D. R. (1990) Crystal structure determinations of synthetic sodium, magnesium, and potassium birnessite using TEM and the Rietveld method: *American Mineralogist* **75**, 477-489.

Rakovan, J. F. and Guggenheim, S. (1991) Rietveld refinement of a IIb-2 clinochlore: in *Proceedings of the 28th Annual Clay Minerals Society Meeting*, 132.

Rautureau, M. and Tchoubar, C. (1974) Précisions concernant l'analyse structurale de la sépiolite par microdiffraction électronique: *Compt. Rendus. Acad. Sc. Paris* **278B**, 25-28.

Rietveld, H. M. (1967) Line profiles of neutron powder-diffraction peaks for structure refinement: *Acta Crystallographica* **22**, 151-152.

Rietveld, H. M. (1969) A profile refinement method for nuclear and magnetic structures: *Journal of Applied Crystallography* **2**, 65-71.

Sen Gupta, P. K., Schlemper, E. O., Johns, W. D., and Ross, F. (1984) Hydrogen positions in dickite: *Clays & Clay Minerals* **32**, 483-485.

Suitch, P. R. and Young, R. A. (1983) Atom positions in highly ordered kaolinite: *Clays & Clay Minerals* **31**, 357-366.

Veblen, D. R. (1985) High-resolution transmission electron microscopy: Chapter 4 in: *Electron Microscopy in the Earth Sciences,* J. C. White, ed., Mineralogical Assoc. Canada, 63-90.

Walker, J. R. and Bish, D. L. (1989) Rietveld refinement of II*b* chlorite: in *Proceedings of the 26th Annual Meeting of the Clay Minerals Society,* 73.

Walker, J. R. and Bish, D. L. (1992) Application of Rietveld refinement techniques to a disordered II*b* Mg-chamosite: *Clays & Clay Minerals* **40**, 319-322.

Wiles, D. B. and Young, R. A. (1981) A new computer program for Rietveld analysis of X-ray powder diffraction patterns: *Journal of Applied Crystallography* **14**, 149-151.

Young, R. A. and Hewat, A. W. (1988) Verification of the triclinic crystal structure of kaolinite: *Clays & Clay Minerals* **36**, 225-232.

Young, R. A. and Wiles, D. B. (1982) Profile shape functions in Rietveld refinements: *Journal of Applied Crystallography* **15**, 430-438.

ILLITE CRYSTALLITE THICKNESS BY X-RAY DIFFRACTION

D. D. Eberl and Alex Blum

CONTENTS

Introduction	124
Factors That Affect XRD Peak Broadening	125
Two XRD Methods For Measuring Crystallite Thickness	128
Scherrer Method	128
Warren-Averbach Method	128
Three Other Methods For Measuring Illite Crystal Thickness	130
Analysis By TEM	130
Analysis By SFM	130
Analysis By Fixed Cation Content	133
Comparisons Between XRD Methods And Other Methods	134
Comparisons Between Mean Particle Thickness Measurements	134
Comparisons Between Thickness Distribution Measurements	139
Calculation Of Illite Properties	139
Calculation Of Crystal Size Distribution From Mean Size	139
Calculation Of Other Properties For Illite	140
Expandability And The Kubler Index	144
Summary And Conclusions	146
Acknowledgments	146
References Cited	147
Appendix: The Warren-Averbach Method	150

ILLITE CRYSTALLITE THICKNESS BY X-RAY DIFFRACTION

D. D. Eberl

U.S. Geological Survey
3215 Marine Street
Boulder, CO 80303

and

Alex Blum

U.S. Geological Survey
Mail Stop 420
345 Middlefield Road
Menlo Park, CA 94025

INTRODUCTION

There are two equally valid ways to describe illite-rich crystals composed of illite and smectite. From the MacEwan crystallite viewpoint (Altaner et al., 1988), such mixed crystals are composed of interlayered illite and smectite layers, and the percentage of each (expandability, or, alternatively, percent illite layers) can be determined readily from X-ray diffraction (XRD) peak positions for glycol-solvated samples (Srodon, 1980, 1984; Moore and Reynolds, 1989). According to the interparticle diffraction concept (Nadeau et al., 1984), illite/smectite (I/S) is composed of discrete "fundamental" illite particles that sorb water on their basal X-Y surfaces. "Smectite layers" are water-rich interfaces that diffract coherently along Z with the adjacent illite particles, thereby producing XRD effects for mixed-layer crystals.

Determination of fundamental illite particle sizes is more significant, crystal chemically and geologically, than is measurement of expandability, as will be discussed below. However, the size of the fundamental illite particles that compose MacEwan crystallites is not as easily determined as is the expandability of the MacEwan crystallites. The most commonly used method to determine illite particle thicknesses is transmission electron microscopy (TEM) of Pt-shadowed specimens (Nadeau and Tait, 1987; Srodon et al., 1992). High resolution transmission electron microscopy (HRTEM) also has been employed to determine fundamental particle thicknesses by measuring directly the number and thickness of illite layers in illite particles (Srodon et al., 1990, 1992). Recently, scanning force microscopy (SFM; also known as atomic force microscopy, AFM) has proven to be a promising technique for making these measurements (Lindgreen et al., 1991; Blum and Eberl, 1992).

These microscopic methods suffer from several problems. For example, many crystals must be measured in a single sample to determine a statistically valid mean thickness, and even more measurements are required to determine a valid distribution of thicknesses. It may take several days to measure a single sample. In addition, microscopic methods currently demand specialized and expensive equipment that requires skilled operators. Consequently, microscopic techniques are not likely to be used routinely in geologic investigations of clay particle sizes. It would be advantageous to develop an XRD technique for measuring particle sizes based on XRD peak broadening, because billions of crystals are measured collectively in a few minutes during an X-ray scan. Mean particle thicknesses and particle thickness distributions then could be extracted from the shapes of selected 00ℓ XRD peaks for illite.

FACTORS THAT AFFECT XRD PEAK BROADENING

The present discussion is concerned with using the 00ℓ series of reflections to measure illite crystallite thickness parallel to c^*. Four factors primarily affect XRD peak breadth (Warren and Averbach, 1953; Barrett and Massalski, 1966; Klug and Alexander, 1974):

1. Instrumental effects, which broaden all of the XRD peaks as a complicated function of diffraction angle;

2. Strains or distortions within crystals, which broaden XRD peaks increasingly as a function of the order of the reflection;

3. Stacking faults or mixed-layering within crystals, which broaden different orders of reflection differently, and often non-symmetrically, depending on the type of stacking fault; and

4. Crystallite size or X-ray scattering domain size, which is the quantity of interest. Crystallite size broadens all peaks equally as a function of the order of the reflection, after the theta-dependent functions have been removed from the diffraction profile.

In order to determine illite crystallite thicknesses by XRD peak broadening, the first three effects need to be separated from the fourth. Instrumental broadening is related to factors such as dispersion of the incident radiation, depth of the diffracting layer within the sample, the slit system used in the incident and diffracted beams, etc. (Barrett and Massalski, 1966). It would be very complicated to remove independently each of these effects from the diffraction profile. However, these effects can be removed collectively by using an instrumental standard. In the case of illite, the best standard is a well-crystallized muscovite, the XRD peaks of which are not broadened significantly by factors 2 through 4 above. Practically, this means analyzing several muscovites, and choosing a muscovite for the instrumental standard that has the narrowest XRD peaks. The choice of a standard is not critical, because clay peaks are so broad that the instrumental contribution to peak breadth generally is small (Brindley, 1980).

The second effect, strain broadening, is related to small variations in the fundamental unit cell repeat distance in a crystallite. In metals, strain could result, for example, from high dislocation densities introduced in an alloy during cold working, with each dislocation deforming the crystal lattice in its vicinity. It seems unlikely that illite particles would be deformed by a similar mechanism. In addition, small strains may be difficult to detect for illite because its $d(001)$ is large compared to that of many metals. However, small, random variations in composition between illite layers, in the amount of octahedral iron for example, may give rise to small differences in layer spacings that could vary about the mean illite spacing of 1.0 nm, as is shown in Figure 1A. Calculations were made using NEWMOD© and NEWMOD3C© computer programs (Reynolds, 1985) to test the effect of variations of ±.01 nm on peak breadth. One-nm illite was interlayered with 0.99 nm to 1.01 nm illite, a likely range of spacings for natural illites (Eberl et al., 1987). No detectable change in the breadth of the illite 002 peak was found for Ca-saturated, air-dried clay,

Eberl and Blum

even if 50 percent of the layers were "strained" in non-swelling illite having X-ray scattering domain sizes ranging from 90 to 100 layers. The calculation demonstrates that the effects of possible strain broadening are so small for the illite 002 reflection that they can be neglected.

The third effect, attributed to stacking faults in the metallurgical literature, is analogous to broadening related to the presence of mixed-layering in clay minerals (Figure 1B). XRD peaks for a clay will be shifted and broadened by a coherently diffracting interlayering phase having a different $d(001)$. The magnitude of this effect will depend on the d-value, quantity and ordering of the interlayering phase. The relative magnitude of the effect can be determined by using the "Q-rule" (Moore and Reynolds, 1989), which is a quantification of Méring's principles (Méring, 1949). The Q-factor for each XRD peak can be calculated for illite by dividing the $d(001)$ of the interlayering phase (e.g. 1.7 nm for glycol-solvated smectite layers) by the $d(001)$ of the illite phase (1.0 nm), and multiplying this ratio by the order of the 00ℓ reflection (1, 2, 3, etc.). Q is the difference between this calculated value and the nearest integer. An XRD peak having a Q-value of 0.5 will exhibit maximum broadening, because XRD peaks for the two phases are far apart. An XRD peak having a Q-value of zero will exhibit minimal broadening for small amounts of mixed-layering, because XRD peaks for the two phases exactly superimpose.

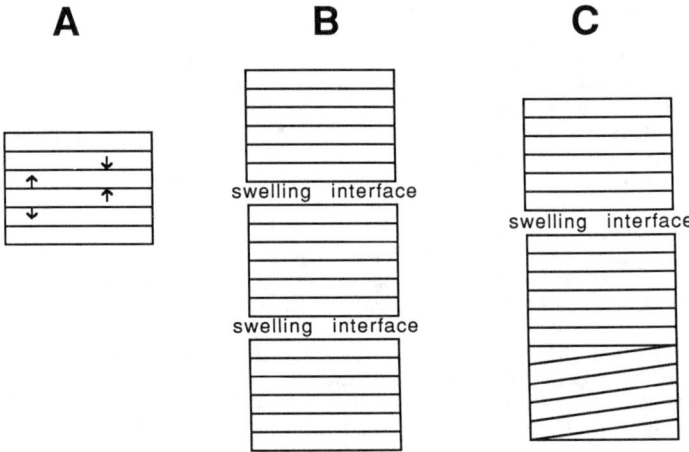

Figure 1. Non-instrumental factors that may lead to XRD peak broadening for illite: **A.** distortions (strains) in the illite lattice; **B.** mixed-layering (stacking faults), related to the presence of swelling surfaces between coherently diffracting fundamental illite particles; and C. an hypothetical defect (tilt boundary; see Barrett and Massalski, 1966, p. 397) within a fundamental illite particle (bottom crystal).

Illite Crystallite Thickness

Q-values have been calculated for the interlayering of several different types of swelling layers in illite (Table 1). Illite has a spacing of 1.0 nm, and the other phases have spacings of 1.25 nm (one water-layer smectite), 1.5 nm (two water-layer smectite), and 1.7 nm (two glycol-layer smectite). It is apparent from Table 1 that the even-order reflections for illite will not be affected by the interlayering of small quantities of 1.5 nm smectite (e.g. Ca- or Sr-saturated smectite). The best peak to use to determine illite crystallite thickness is the 002, because the 004 illite reflection often is weak and asymmetric. Calculations using NEWMOD confirm that this 002 peak is minimally affected by mixed-layering: it changes width at half height by only 0.02° two-theta for I/S having expandabilities ranging from 0% to 20%, using a mean defect free distance of 20 layers and a high N of 50 in the calculation, whereas the 005 peak broadens and then splits into two peaks over this same range in expandability.

Table 1. Q-factors for illite interlayered with various swelling layers.

Reflection	1.25 nm	1.5 nm	1.7 nm
001	0.25	0.5	0.3
002	0.5	0	0.4
003	0.25	0.5	0.1
004	0.5	0	0.2
005	0.25	0.5	0.5

After removal of instrumental broadening (Stokes, 1948) and theta-dependent functions (see Appendix) from the 002 reflection for Ca- or Sr-saturated, air-dried samples, XRD peak breadth results primarily from factor 4 above, the X-ray scattering domain size distribution parallel to c^*. X-ray scattering domain size, which is equivalent to crystallite size, may or may not be equal to fundamental particle thickness, as is shown in Figure 1C. The upper crystal in this figure is defect-free, and therefore particle thickness and X-ray scattering domain size are equal. The lower crystal has a hypothetical tilt defect that causes the crystal to diffract as two separate regions, and crystal thickness is about twice the possible X-ray scattering domain size for this fundamental particle. In addition, if the upper crystal and the upper half of the lower crystal diffract coherently, then X-ray domain thickness can be larger than fundamental particle thickness. One can not readily distinguish by XRD alone whether a measured X-ray scattering domain size is equivalent to fundamental particle size. This problem is inherent in any XRD method for particle size determination.

A second problem inherent to the method is that crystallite sizes greater than about 100 nm will not contribute significantly to XRD peak broadening, and therefore will not influence the measurement of crystallite size. Generally, the mean domain size needs to be about 40 nm or less to provide a meaningful measurement (Siemens, 1990). Therefore, the method generally is restricted to measuring clay crystallite thicknesses, because other clay crystal dimensions are too large. The lower limit for the measurement is slightly larger than 2 nm, which means that the I/S must be less than about 20% expandable.

TWO XRD METHODS FOR MEASURING CRYSTALLITE THICKNESS

The next step is to analyze the 002 peak for Ca-saturated, air-dried illite to determine mean X-ray scattering domain thickness and domain thickness distribution. Two methods will be considered: the Scherrer equation, and the method of Warren and Averbach.

Scherrer Method

The Scherrer equation, given by Klug and Alexander (1974), is:

$$L = K\lambda / B(\cos\theta), \qquad (1)$$

where L = the volume average crystallite thickness in Å along c^*; K is a constant = 0.91 for clays (Brindley, 1980); θ = two-theta diffraction angle/2; λ = wavelength of radiation used (1.541838 Å for Cu Kα); and B = the corrected breadth of the XRD peak at half height measured in radians (degrees/57.3 = radians) on the two-theta scale. The observed breadth is corrected as follows:

$$B = \sqrt{\beta^2 - b^2}, \qquad (2)$$

where β is the measured breadth at half height for the sample's peak, and b is the breadth at half height for that of the instrumental standard (Brindley, 1980). The Scherrer equation will give a realistic mean thickness of the coherent scattering domain only when the XRD peak has not been broadened by other factors such as strain and mixed-layering.

Warren-Averbach Method

The Warren-Averbach (W-A) method (see Appendix) can separate strain broadening from domain size broadening through a Fourier analysis of at least two XRD peaks that are related by n in Bragg's Law (Warren and Averbach, 1950, 1953; Warren, 1959; Barrett and Massalski, 1966; Klug and Alexander, 1974). Because domain size broadening of XRD peaks does not change as a function of diffraction order (after theta-dependent functions have been removed from the diffraction profile -- see Eq. 21 in the Appendix), whereas strain broadening increases as a function of diffraction order, an extrapolation of the Fourier coefficients used to model two or more 00ℓ XRD peaks to the zeroth order reflection removes the strain contribution. Therefore, mean domain thickness and domain thickness distributions can be determined from the corrected coefficients as outlined in the Appendix.

Strain broadening for the Ca-saturated illite 002 reflection is unimportant, and this peak is unaffected by swelling. Therefore, Fourier analysis of this single peak can yield crystallite thicknesses. An 00ℓ XRD peak for illite can be considered to be the sum of diffraction intensities resulting from X-ray scattering domains having different thicknesses -- the coarser the crystallite, the sharper its contribution to the XRD peak. Figure 2 shows how a calculated XRD peak can be the sum of three different crystallite thicknesses, in this case occurring in equal proportions and having 10, 14 and 18 coherently diffracting illite layers each. Fourier analysis reverses the summation process, and extracts the contribution of each domain thickness to the diffraction intensity by modeling the contributions of the various thicknesses as cosine waves.

Illite Crystallite Thickness

The Siemens Corporation D5000[1] software performs the Warren-Averbach crystallite size analysis automatically. Analysis proceeds as follows. First, Ca- or Sr-saturated illite is prepared as an oriented XRD mount. It is analyzed by XRD in the air-dried state using incident and diffracted beam Soller slits from 15 to 20 °2θ using 0.02 °2θ steps with one or two seconds count time per step. The 002 illite peak then is modeled, using the Siemens' program FIT, with a split Pearson VII function using separate exponents because this function creates the most precise mathematical model of the peak's shape. Interfering peaks also are modeled at this time so that their contribution to the 002 illite peak can be removed, and the Kα2 contribution is separated from the Kα1 peak. An instrumental standard also is analyzed and its peak modeled in exactly the same manner. Often peak fitting needs to be repeated two or three times, with modified starting exponents or background intensities, to insure that the tails of the peaks are well-fit by the split Pearson function.

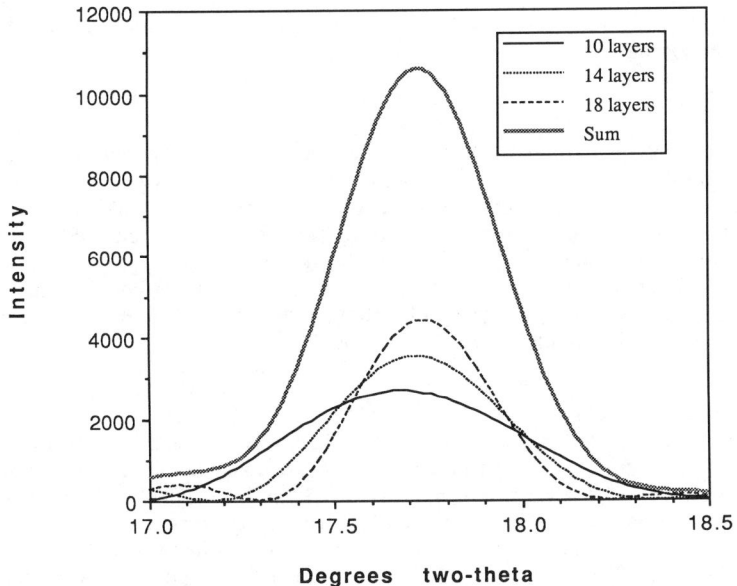

Figure 2. XRD peaks for Ca-saturated, air-dried I/S having 5% smectite layers, calculated for various particle thicknesses using NEWMOD© (Reynolds, 1985) The sum of diffraction intensities from various crystallite sizes yields the observed XRD peak.

---

[1]The use of brand names is for identification purposes only and does not constitute endorsement by the U. S. Geological Survey.

Next the CRYSIZ program is opened, and the Warren-Averbach method is executed using sample and instrumental standard inputs from the files created by the FIT program. The program then prints out the mean crystallite thickness and the crystallite thickness distribution, calculated as outlined in the Appendix. In some W-A analyses the mean crystallite size that is printed out and the mean crystallite size that can be calculated from the crystallite size distribution are not exactly equal. This situation is especially common for samples dominated by very thin crystallites, and in this case the latter value seems to be the more accurate, based on comparisons with other techniques. The problem may result from inaccuracies in extrapolation of the initial slope of the curve A_L^S versus L to $A_L^S = 0$ for very thin crystallites (see Appendix and Figure A2).

THREE OTHER METHODS FOR MEASURING ILLITE CRYSTAL THICKNESS

Illite crystal thicknesses can be determined by transmission electron microscopy (TEM) on Pt-shadowed crystals (Nadeau and Tait, 1987), and by scanning force microscopy (SFM, Blum and Eberl, 1992). In addition, fixed cation content can be used to calculate mean fundamental particle thickness (Srodon et al., 1992). Crystal thicknesses measured by these methods can be compared with crystallite thicknesses measured by the Scherrer equation and by the Warren-Averbach method for the same samples.

Analysis by TEM

Pt-shadowed crystals of Zempleni illite are shown in Figure 3. The "shadows" appear as bright areas next to the crystals. Crystal thicknesses can be calculated from the tangent of the shadowing angle by measuring the length of the shadow. Uncorrected mean thicknesses ($\overline{T}_{TEM}$) calculated from these measurements then are corrected for differences in measured X-Y areas for the particles, and the corrected mean ($\overline{T}_{cTEM}$) is calculated (Srodon et al., 1992). This correction is necessary if there is a relation between particle area and particle thickness, because then the diffracting mass of the particles will change as a function of thickness. This correction generally is small for thicknesses less than 20 nm (Table 2).

Analysis by SFM

Similar reasoning applies to SFM measurements of illite thicknesses to determine the mean and the corrected mean ($\overline{T}_{SFM}$ and $\overline{T}_{cSFM}$). SFM measurements are considerably quicker and easier to perform than are TEM Pt-shadow measurements (Blum and Eberl, 1992). Moreover, SFM reveals additional details of crystal surface morphologies (Figure 4). Measurements were made using suspensions of the clay that had been ultra-sonically dispersed in distilled water on freshly cleaved muscovite surfaces. The microscope was a Digital Instruments SFM, primarily a Nanoscope III, in height mode, with no filters, a scan rate of ~5kHz, and moderate gains of 3. SFM can accurately measure the topography of surfaces with a resolution that ideally approaches ±0.01 nm. However, measurement of 1.0 nm cleavage steps on muscovite yielded an estimated uncertainty of ±0.05 nm. During the routine measurement of clay particle thicknesses, the estimated accuracy is the greater of ±0.2 nm or ±10%. The larger uncertainty primarily results from the small lateral extent of most clay particles, and from the presence of multiple particles in the image, both of which exacerbate problems with accurate background flattening, edge effects and instrumental noise.

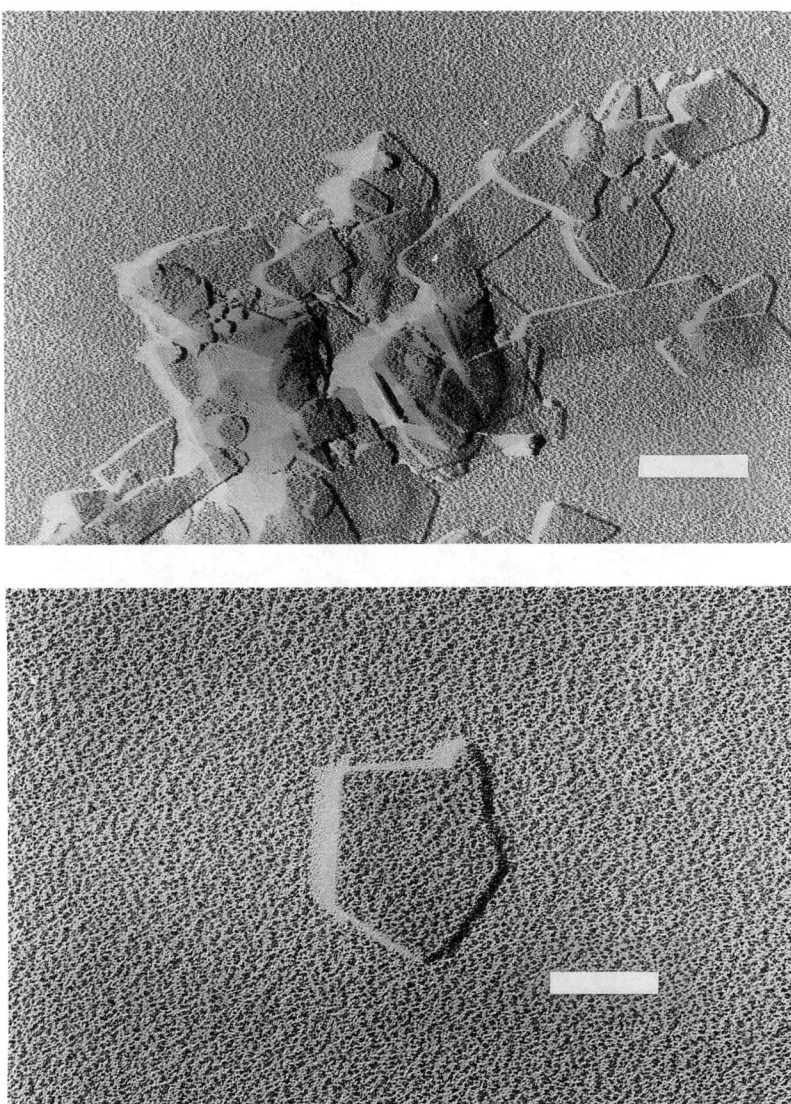

Figure 3. TEM images of Zempleni illite. Top image shows overlapping particles which may give rise to interparticle diffraction effects. Bottom image shows a single Pt-shadowed fundamental illite particle. Horizontal bars are 1 μm. Images courtesy of Jan Srodon.

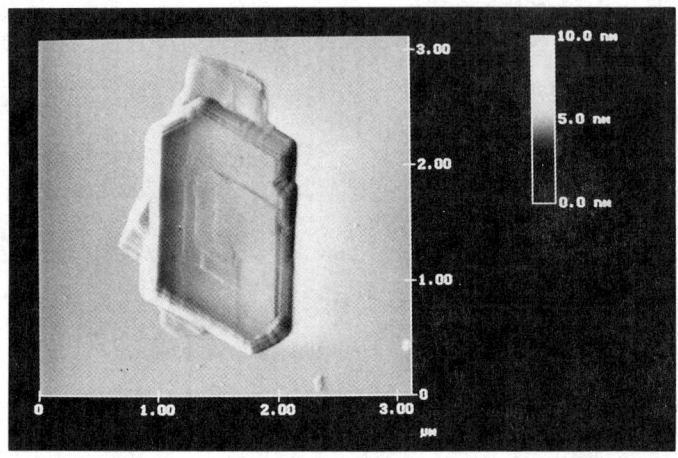

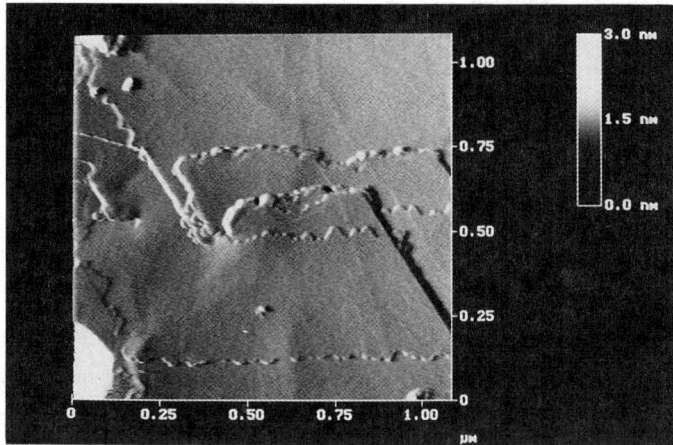

Figure 4. SFM images of AR1 illite. Top image shows a growth spiral. Such spirals are present on several percent of the larger grains. Bottom image shows a more detailed image of a growth spiral on another crystal. The images are stored as a digital surface with both lateral and vertical coordinates, and therefore particle dimensions are readily accessible.

Illite Crystallite Thickness

Table 2. Mean sample thickness $(\overline{T})$ analyzed by several techniques.[1]

Sample (ref.)	$\overline{T}_{FIXED}$	$\overline{T}_{TEM}$	$\overline{T}_{cTEM}$	$\overline{T}_{SFM}$	$\overline{T}_{cSFM}$	$\overline{T}_{W-A}$	$\overline{T}_{Sch}$
AR1 (1, 8)	44.5	n.a.	n.a.	30.8	40.8	21.8	21.9
AR1R (1, 8)	29.7	n.a.	n.a.	n.a.	n.a.	12.2	13.5
KALK. (6)	2.54	n.a.	n.a.	n.a.	n.a.	4.5	6.7
KAUBE (3, 4)	n.a.	34.4	52.9	n.a.	n.a.	22.6	26.2
LF7 (1, 8)	8.9	n.a.	n.a.	n.a.	n.a.	7.7	10.3
LF8 (1, 8)	n.a.	n.a.	n.a.	8.7	10.1	7.9	7.7
LF10 (1, 8)	6.8	n.a.	n.a.	n.a.	n.a.	8.3	9.1
M4 (3, 5)	2.6	2.7	2.6	n.a.	n.a.	2.5	6.3
M11 (3, 5)	5.2	4.8	5.4	n.a.	n.a.	7.0	8.6
MF987 (2)	n.a.	n.a.	n.a.	22.1	41.8	38.9	43.4
MF998 (2)	n.a.	n.a.	n.a.	33.1	70.8	38.6	43.7
RM3 (1, 8)	12.7	n.a.	n.a.	n.a.	n.a.	10.9	15.1
RM4 (1, 8)	8.1	n.a.	n.a.	n.a.	n.a.	n.a.	11.7
RM5 (1, 8)	11.1	n.a.	n.a.	n.a.	n.a.	6.3	9.8
RM6 (1, 8)	8.1	n.a.	n.a.	n.a.	n.a.	7.3	10.3
RM8 (1, 3, 8)	6.4	5.8	5.8	n.a.	n.a.	5.0	8.5
RM12 (1, 8)	7.4	n.a.	n.a.	n.a.	n.a.	7.1	7.2
RM13 (1, 8)	8.9	n.a.	n.a.	n.a.	n.a.	7.5	8.2
RM22 (1, 8)	n.a.	n.a.	n.a.	11.1	11.9	5.0	6.9
RM28 (1, 8)	9.9	n.a.	n.a.	n.a.	n.a.	9.0	11.6
RM30 (1,3, 8)	9.9	10.6	11.6	10.6	10.6	10.0	8.1
RM35A (1, 3, 8)	5.6	5.6	6.9	n.a.	n.a.	5.1	12.1
RM35D (1, 8)	3.9	n.a.	n.a.	6.2	6.9	3.6	6.5
SG1 (1, 8)	11.1	n.a.	n.a.	n.a.	n.a.	19.1	24.9
SG4 (1, 3, 8)	17.8	14	17.8	n.a.	n.a.	16.7	22.3
SLV.HILL (4)	4.5	n.a.	n.a.	n.a.	n.a.	3.1	7.4
T9 <0.2 (3, 5)	2.7	2.6	2.7	n.a.	n.a.	2.7	n.a.
ZEMP. (3)	4.5	3.2	3.2	3.6	3.6	3.1	4.5

[1] Samples analyzed were the <1 or <2 µm size fraction, unless otherwise noted. All XRD data were collected from Sr-saturated, air-dried samples. $\overline{T}_{FIXED}$ = mean thicknesses (in nm) calculated from fixed cation content (see Eq. 3); $\overline{T}_{TEM}$ = mean thickness (in nm) measured by TEM and Pt-shadowing; $\overline{T}_{cTEM}$ = mean TEM thicknesses (in nm) corrected for particle X-Y areas; $\overline{T}_{SFM}$ = mean thicknesses (in nm) measured by SFM; $\overline{T}_{cSFM}$ = mean SFM thicknesses (in nm) corrected for the X-Y areas of the particles; $\overline{T}_{W-A}$ = mean thicknesses (in nm) measured by the Warren-Averbach method for the 002 reflection for air-dried samples; $\overline{T}_{Sch}$ = mean thicknesses (in nm) measured by the Scherrer equation (Eq. 1); n.a. = not analyzed. Information concerning the samples is in references (ref.): (1) Eberl et al. (1987); (2) Hunziker et al. (1986); (3) Srodon et al. (1992); (4) Srodon and Eberl (1984); (5) Srodon et al. (1986); (6) Hower and Mowatt (1966); (7) Güven (1972); (8) Eberl and Srodon (1988).

Analysis by Fixed Cation Content

A third method for determining mean illite crystal thicknesses relies on the observation that the fixed cation (K + Na) content of illite layers is constant at about 0.89 equivalents per $O_{10}(OH)_2$ (Srodon et al., 1992). Therefore a 2.0 nm illite fundamental particle containing one fixed interlayer would have a fixed cation content of 0.89 - (0.89/2) = 0.45 equivalents, because half of its interlayers are located on particle surfaces, and therefore this (K + Na) would be exchangeable.

Eberl and Blum

Likewise, a 3.0 nm thick illite particle having two fixed interlayers would have 0.59 equivalents of fixed cations, and so on. The equation for calculating the mean thickness ($\overline{T}$, in nm) of fundamental illite particles in a sample based on the equivalents of fixed cations (FIX) per $O_{10}(OH)_2$ is (Srodon et al., 1992):

$$\overline{T} = \left(1 - \frac{FIX}{0.89}\right)^{-1}. \tag{3}$$

COMPARISONS BETWEEN XRD METHODS AND OTHER METHODS

Comparisons between Mean Particle Thickness Measurements

Illite thicknesses measured by the microscopic, the fixed cation, and the two XRD methods are compared in Tables 2 and 3, and in Figures 5 through 8. The microscopic (TEM + SFM) methods and the fixed cation method Eq. (3) give comparable results (Figure 5; $r^2 = 0.97$), as has been shown previously (Srodon et al., 1992; Blum and Eberl, 1992). The microscopic plus fixed cation methods and the W-A method also give comparable results, up to a fundamental particle thickness of about 20 nm (Figure 6A; $r^2 = 0.71$). This correlation suggests that the X-ray scattering domain thickness (Figure 2) that is measured by the W-A technique is approximately equivalent to the thickness of fundamental illite particles. At thicknesses larger than 20 nm, however, crystallite thicknesses measured by the W-A method generally are smaller than fundamental particle thicknesses measured by the other techniques (Figure 6B).

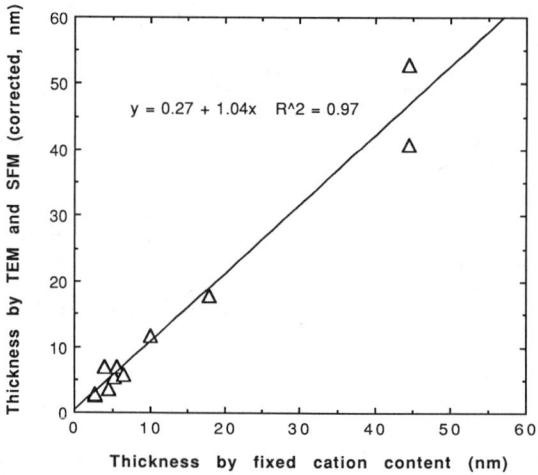

Figure 5. Comparison between illite particle thicknesses measured by TEM + SFM with those measured by the fixed cation method.

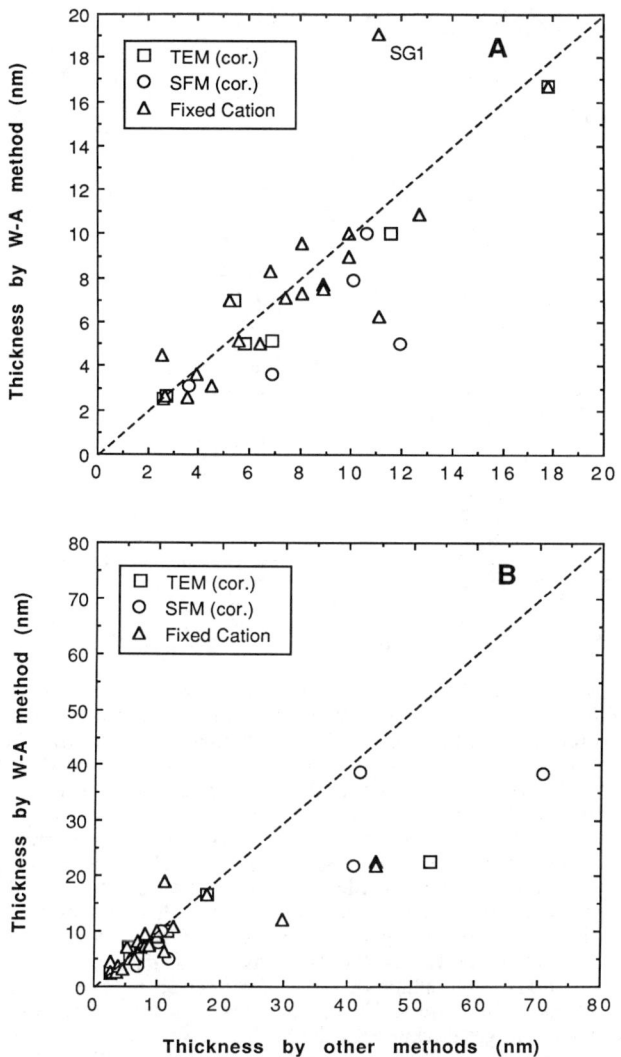

Figure 6. Comparison between fundamental illite particle thicknesses measured by the TEM + SFM + fixed cation methods with crystallite thicknesses measured by the Warren-Averbach method: **A**. thicknesses ranging from 2 to 20 nm; **B**. the complete data set.

Table 3. Samples analyzed only by Scherrer and W-A techniques.[1]

Sample (ref. no.)	$\overline{T}_{W-A}$	$\overline{T}_{Sch}$
BRM11	6.5	12.2
GC1	5.2	7.4
KA545	2.6	8.9
L2A1,>1 (4)	20.9	24.2
MBLHD (7)	2.6	6.3
MF4 (2)	2.8	20.2
MF22 (2)	2	15.7
MF23 (2)	9.4	16.3
MF34 (2)	9.9	24.1
MF53 (2)	18.1	32.6
MF54 (2)	23.1	41.2
MF59 (2)	8.2	19.3
MF60 (2)	21.8	26.0
MF71 (2)	21.0	27.3
MF65 (2)	14.1	23.3
MF74 (2)	26.8	35.1
MF82 (2)	30.8	38.7
MF341 (2)	9.6	26.0
MF361 (2)	42.2	59.6
MF664 (2)	16.8	32.6
MF731 (2)	26.7	38.0
MF1170 (2)	20.1	29.9
MF1324 (2)	5.4	14.5

[1] See Table 2 for key.

The microscopic and fixed cation methods most likely give more accurate mean fundamental particle thicknesses than does the W-A technique. The smaller values found by the W-A technique for particles having thicknesses greater than about 20 nm may be related to the presence of defects in the crystals (Figure 1C, lower crystal) that cause the coherent X-ray scattering domains in the Z direction to be smaller than the crystal thicknesses, to the presence of large crystals that do not broaden the XRD peaks, to the presence of swelling interlayers that deviate significantly from 1.5 nm, and to the presence of mixtures of two or more types of illite having different $d(001)$. Defects become more probable as crystals become thicker (Ergun, 1970). Large crystals also become more common as the mean size increases. Low-grade metamorphic terrains may contain both detrital and authigenic illite having different d-values. From such reasoning, X-ray methods could commonly give values smaller than thicknesses measured by the microscopic and fixed cation methods. X-ray methods could give values larger than those measured by the other methods if the shape of the 002 reflection is related to the presence of coherently diffracting stacks of fundamental illite particles. Only one of the samples, SG1 in Figure 6A, gave a significantly larger thickness by the W-A method.

Thicknesses calculated by the Scherrer equation (Eq. 1) are larger than those calculated by the W-A method (Tables 2 and 3; Figure 7A, with $r^2 = 0.85$). The Scherrer equation gives much more scattered data than does the W-A method when compared to the microscopic and fixed cation methods. A linear fit of the Scherrer data in Figure 7B gives $r^2 = 0.28$ compared to $r^2 = 0.71$ for the W-A data in Figure 6A. Therefore, the W-A method is the preferred method. Another advantage to the W-A method is that it yields crystallite thickness distributions.

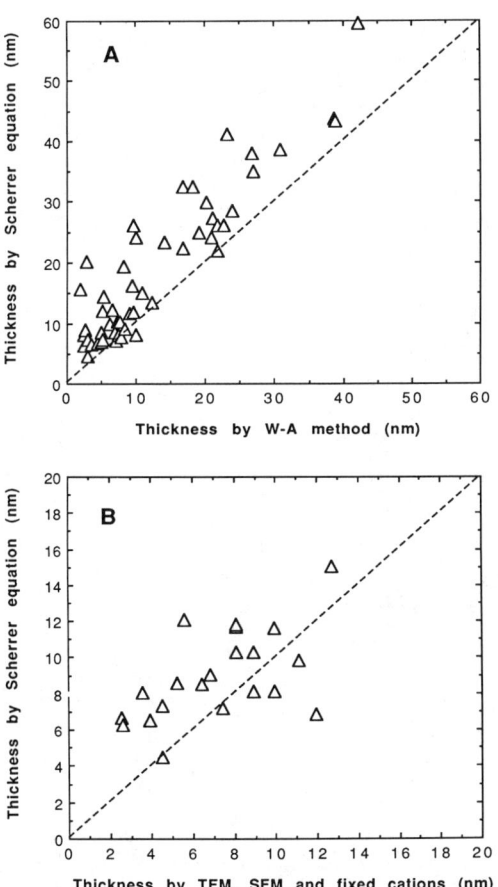

Figure 7. Comparison between illite crystallite thicknesses measured by the Scherrer method with thicknesses measured by other methods: **A**. Scherrer versus Warren-Averbach method; **B**. Scherrer versus TEM + SFM + fixed cation methods.

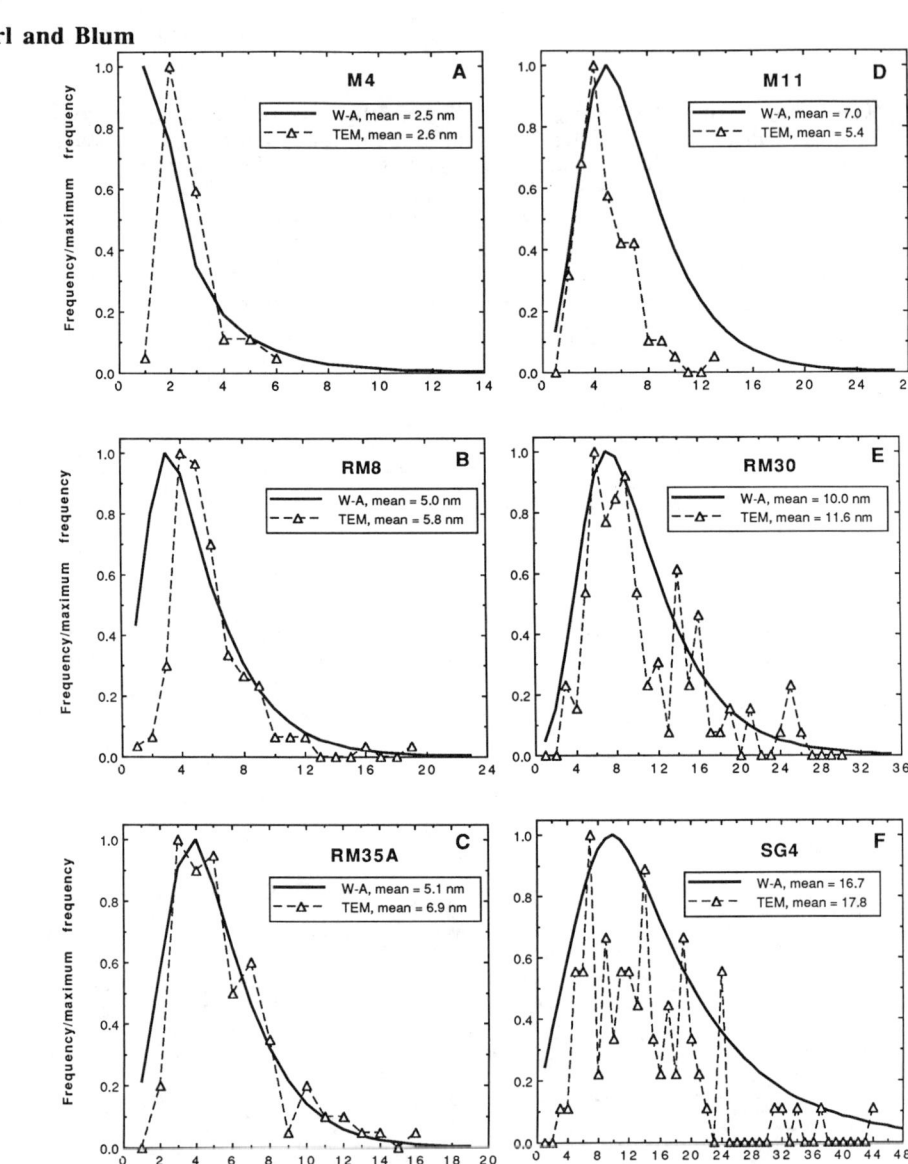

Figure 8. Fundamental illite particle thickness distributions measured by TEM (dashed lines) compared with crystallite thicknesses measured by the Warren-Averbach (W-A) method (solid lines). TEM data courtesy of Jan Srodon.

Illite Crystallite Thickness

Comparisons between Thickness Distribution Measurements

Comparisons between thickness distributions measured by the W-A method and by TEM (Figure 8) indicate that the methods give comparable results, thereby offering additional evidence that the scattering domain measured by the W-A method is equivalent to the thickness of fundamental particles. In order to compare data between the two methods, frequency was plotted as frequency divided by maximum frequency. This manipulation was necessary because the TEM data are noisy, probably because the TEM measurements are based on about 100 observations for each sample, whereas the Warren-Averbach measurement averages all of the diffracting crystals in the X-ray beam. The TEM data become noisier as the mean particle thickness increases from Figure 8A through Figure 8F, because the particle size distribution spreads out (Table 3 in Nadeau, 1987), an effect probably related to crystal growth by Ostwald ripening (Baronnet, 1982; Eberl and Srodon, 1988; Eberl et al., 1990). Therefore, the larger the mean thickness, the more TEM and SFM measurements are required to determine an accurate distribution.

CALCULATION OF ILLITE PROPERTIES

Calculation of crystal size distribution from mean size

Much can be inferred about the properties of an illite from measurement of its mean fundamental particle thickness ($\overline{T}$). For example, one can calculate the thickness distribution for many samples from the equation for a log-normal distribution (Eberl et al., 1990):

$$f(\omega) = \left[\frac{1}{\omega\beta(2\pi)^{\frac{1}{2}}}\right] \cdot \exp\left\{-\left(\frac{1}{2\beta^2}\right)[\ln(\omega) - \alpha]^2\right\}, \tag{4}$$

where $f(\omega)$ = frequency of ω, ω = thickness/mean thickness = $T/\overline{T}$, $\beta^2 = 0.29$, and $\alpha = -0.133$. The values for parameters β^2 and α were calculated from a reduced particle thickness distribution measured by the W-A method for sample RM30, and have been corrected slightly from those given in Eberl et al. (1990). Eq. (4) has been solved in Table 4 for a range of $T/\overline{T}$ for the example $\overline{T}$ = 10 nm. Particle thicknesses are calculated as a function of frequency by multiplying the first column in Table 4 by the measured mean particle thickness.

Use of Eq. (4) to calculate particle size distributions is based on the observation that the particle sizes of most clays, and that of many other minerals, follow the same reduced particle size distribution, as is shown for illites measured by the W-A method (Figure 9A), illites measured by other methods (Figure 9B), other clay minerals (Figure 9C), and metamorphic and other minerals (Figure 9D). Exceptionally thin fundamental illite particles may deviate from this general rule (see Figure 3 in Eberl et al., 1990). The particle size distribution is reduced by plotting frequency/maximum frequency versus size/mean size (Baronnet, 1982). The log-normal reduced plot has been found to be independent of measurement technique, because particle dimensions measured for a variety of minerals by XRD, scanning electron microscopy (SEM), TEM, light microscope and field flow fractionation-mass spectrometry (FFF-MS) all follow the same relationship given in Eq. (4). Equation (4) is independent of the particular crystal dimension measured, provided the measurements are made consistently of the same dimension. Other minerals found to fit this master curve are cherts from the deep sea drilling project (unpublished data using W-A method), chlorites from the North Sea basin (SEM data of Jahren, 1991), illites

Table 4. Solutions to Eq. (4) in text for the example $\overline{T}$ = 10 nm.

$T/\overline{T}$	$f(\omega)$	T
0.2	0.088	2
0.4	0.661	4
0.6	0.994	6
0.7	1.0	7
0.8	0.941	8
1.0	0.74	10
1.2	0.536	12
1.4	0.372	14
1.6	0.254	16
1.8	0.173	18
2.0	0.117	20
2.2	0.080	22
2.4	0.055	24
2.6	0.038	26
2.8	0.026	28
3.0	0.019	30
3.2	0.013	32
3.4	0.009	34
3.6	0.007	36
3.8	0.005	38
4.0	0.004	40

from the Paris Basin (TEM data of Lanson and Champion, 1991), and Holocene dolomites from a peritidal environment in Belize (SEM data of Gregg et al., 1992).

"Steady state" plots, such as those shown in Figure 9, are indicative of Ostwald ripening, whereby smaller, dissolving crystals contribute matter to larger, growing crystals of the same phase. The smaller crystals are less stable because they have a larger specific surface area. The shape of the reduced distribution remains steady during ripening despite an increase in mean particle size. The shape of the reduced distribution is indicative of a rate controlling step during the ripening process (Baronnet, 1982), but the log-normal distribution shown in Figure 9, which is the common steady state distribution in nature, has not yet been related to a particular ripening mechanism (Eberl et al., 1990).

Calculation of Other Properties for Illite

Due at least partly to the regular relation between mean particle size and particle size distribution (Eq. 4), several other crystal-chemical relations for illite can be calculated from a knowledge of mean thickness ($\overline{T}$, in nm). This remarkably regular set of relations, taken from Nadeau (1985, 1987) and Srodon et al. (1992), have been recast where necessary in terms of mean thickness ($\overline{T}$) rather than in terms of maximum expandability, where maximum expandability = $100/\overline{T}$. The fixed cation content (*FIX*, in equivalents per $O_{10}(OH)_2$) is calculated from $\overline{T}$ by rearranging the theoretical Eq. (3) (Srodon et al., 1992; $r^2 = 0.94$ for the experimental equation):

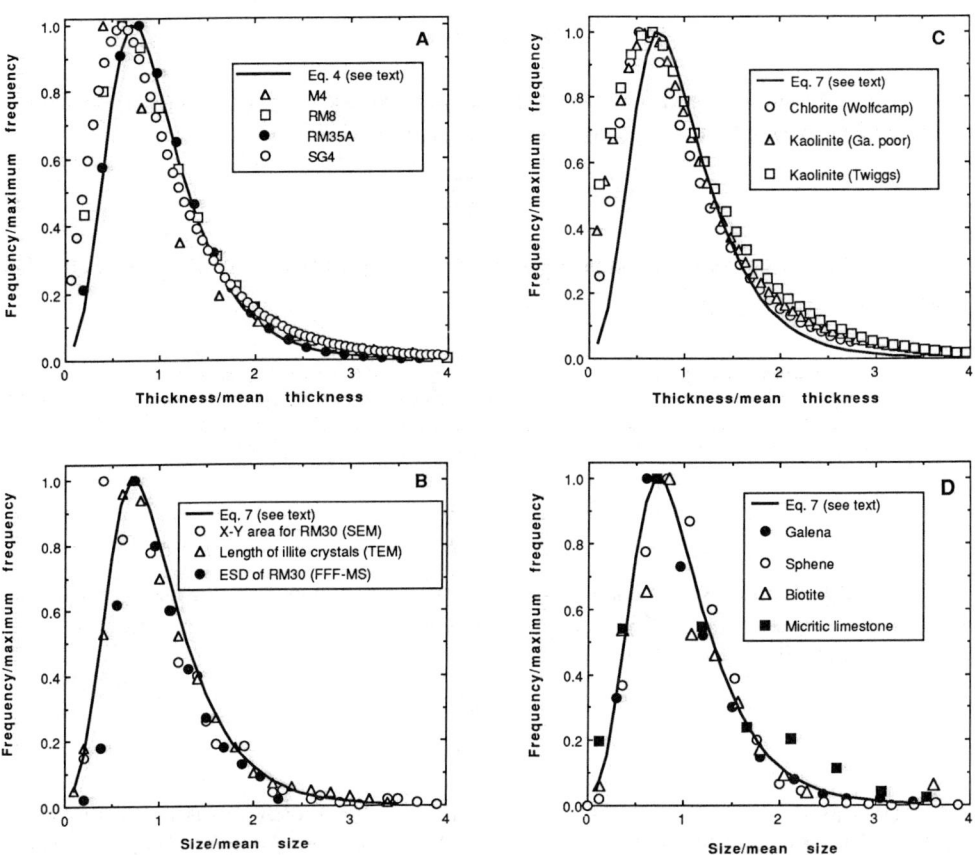

Figure 9. Reduced plots of crystal sizes for various minerals (individual symbols) compared with the predicted function calculated from Eq. (4) in the text (solid lines): **A**. reduced thickness distributions for illites measured by the Warren-Averbach technique (this paper); **B**. reduced distributions for illite X-Y areas (from Eberl et al., 1990) measured by SEM, for illite crystal lengths (from Inoue et al., 1988) measured by TEM, and for illite equivalent spherical diameters (ESD) measured by field flow fractionation-mass spectrometry (unpublished data of Howard Taylor, USGS); **C**. reduced distributions for clays measured by the Warren-Averbach technique (from Eberl et al., 1990); **D**. reduced distributions for other minerals measured by the petrographic microscope (references in Eberl et al., 1990).

$$FIX = \frac{0.89(\overline{T}-1)}{\overline{T}}. \qquad (5)$$

An empirical relationship between the surface area measured by the ethylene glycol monoethyl ether (EGME) method (SA, in m^2/g) and $\overline{T}$ is (Srodon et al., 1992):

$$SA = \frac{833}{\overline{T}}. \qquad (6)$$

Equation (6) agrees well with the prior equation given by Nadeau (1985) who measured illite surface areas by TEM: $SA = \frac{825}{\overline{T}}$.

The cation exchange capacity (CEC, in milliequivalents per 100 g) is calculated from the following equation which assumes a charge of -0.4 equivalents/$O_{10}(OH)_2$ for the smectite surfaces (a reasonable assumption based on the data of Srodon et al., 1992), where M is the formula weight, usually about 375, for the clay per $O_{10}(OH)_2$ (Srodon et al., 1992):

$$CEC = \frac{4.2 \times 10^4}{\overline{T}M}. \qquad (7)$$

The mean X-Y illite particle area, which is defined as the product of the mean particle length and the mean particle width, can be calculated by solving the following empirical equation for $\overline{A}$ (in nm^2; Nadeau, 1987; $r^2 = 0.985$):

$$\overline{T} = \frac{(\overline{A})^{0.571}}{127.9}. \qquad (8)$$

Additionally, as $\overline{A}$ and $\overline{T}$ increase, their distribution becomes broader, as is demonstrated by empirical relations found between these parameters and their standard deviations ($Std_{\overline{A}}$ and $Std_{\overline{T}}$). According to Table 2 in Nadeau (1987):

$$Std_{\overline{A}} = \frac{0.94\overline{A}}{10^4} + 3.18, \text{ and} \qquad (9)$$

$$Std_{\overline{T}} = 0.75\overline{T} + 0.54, \qquad (10)$$

with $r^2 = 0.97$ and 0.87, respectively.

Other interesting relations from Nadeau (1987) are those between illite mean particle volume ($\overline{V}$, in nm^3) and $\overline{A}$ ($r^2 = 0.98$):

Illite Crystallite Thickness

$$\text{Log}\overline{V} = 1.53\log\left(\frac{\overline{A}}{10^4}\right) + 4.21, \tag{11}$$

and between $\overline{V}$ and mean particle mass ($\overline{M}$, in g/particle):

$$\overline{M} = \overline{V}\rho 10^{-21}, \tag{12}$$

where ρ is the density (about 2.6) in g/cm^3. The number of particles per gram (N_g) then is given by:

$$N_g = \frac{1}{\overline{M}}. \tag{13}$$

Eqs. (11), (12) and (13) indicate that a sample such as RM30, which has a mean particle area of 22.97 x 10^4 nm^2 (Nadeau, 1987), contains about 2 x 10^{14} (200 trillion) particles per gram. It is remarkable that measurement, for example, of about a hundred particle thicknesses by TEM or SFM (Table 2) can be representative of the whole sample.

Nadeau (1987) also gives equations for calculating the basal surface area (S_b), the edge on lateral surface area (S_l) and the total specific surface area (S) in m^2/g:

$$S_b = 2(\overline{A})10^{-22}N_g, \tag{14}$$

$$S_l = (\overline{T})(\overline{P})10^{-18}N_g, \tag{15}$$

$$S = S_b + S_l, \tag{16}$$

where $\overline{P}$ is the mean perimeter of a regular hexagon given by $\overline{P} = 3.72\left(\frac{\overline{A}}{10^4}\right)^{0.5}$.

A final empirical relationship can be extracted from the data of Srodon et al. (1986) and Eberl et al. (1987) by plotting the expandability (X) of mixed-layer I/S versus $\overline{T}$ measured for the same samples by the fixed cation method:

$$\text{Log}\overline{T} \approx 1.66 - 1.12(\log X) + 0.15(\log X)^2, \tag{17}$$

with $r^2 = 0.94$. If there is no other recourse, one could measure XRD expandability for I/S (Srodon, 1980, 1984; Moore and Reynolds, 1989), solve Eq. (17) for the mean thickness, and use this thickness to solve the other equations. However it would be much better to measure mean particle thickness directly, because the interpretation of XRD expandability measurements is subject to several problems, which will be discussed in the next section.

EXPANDABILITY AND THE KUBLER INDEX

Two measurements commonly used to follow illite diagenesis and metamorphism are expandability and the Kubler "crystallinity" index. Expandability, which is a measurement of the percentage of swelling interlayers in I/S, is most useful when studying I/S samples from the lowest temperature alteration regimes. The Kubler index, which is the width at half height of the illite 001 XRD reflection, is most useful for studying illites that have been subjected to higher temperatures, after most of the expandability has disappeared.

Diagenetic and metamorphic reactions involving illite also may be followed by measuring particle size, or particle thickness. These measurements are more closely related to the thermodynamic status of the sample system, because during the metamorphism of illite there is a tendency to minimize the free energy (ΔG_{total}) of illite crystals, where:

$$\Delta G_{total} = \Delta G_{\infty\ crystal} + \Delta G_{surface}, \qquad (18)$$

and $\Delta G_{surface} = \Sigma\ \sigma_i S_i$, where σ_i = the surface free energy and S_i = the specific surface area. The free energy of the interior of illite crystals ($\Delta G_{\infty\ crystal}$) can be minimized through the annealing of defects and the elimination of stacking faults and compositional impurities within the fundamental illite particles. Total crystal free energy also can be lowered by reducing the specific surface area in the second term through illite crystal growth, thereby decreasing expandability. Thus illite particle surface area may be a good parameter to determine the degree of illite metamorphism. As was discussed previously, illite surface area is related to mean illite particle thickness by Eq. (6). XRD expandability and the Kubler index, however, are less direct measures of the extent of reaction.

Expandability (X) can be defined as:

$$X = \text{number of expanding interlayers/total number of interlayers.} \qquad (19)$$

From this definition, expandability will depend both on the thickness of fundamental illite particles composing MacEwan crystallites, and on the number of fundamental illite particles in MacEwan crystallites. Figure 10A shows the effect of changing fundamental particle thickness on expandability for two MacEwan crystallites, keeping the number of fundamental illite particles constant at two per stack. The fundamental illite particles in the crystallite on the left are composed of one illitic interlayer each, and, from Eq. (19), expandability = 33%. The crystallite on the right is composed of particles having three illitic interlayers each, and expandability = 14%. Figure 10B shows the effect of changing the number of illite particles in MacEwan crystallites on expandability for elementary illite particles composed of one illite interlayer each: one such illite particle is 0% expandable; two in a stack give 33%; and three in a stack yield 40% expandable. The maximum expandability for the situation depicted in Figure 10B would be 50% for a MacEwan crystallite that is composed of an infinite number of such elementary illite particles.

Illite Crystallite Thickness

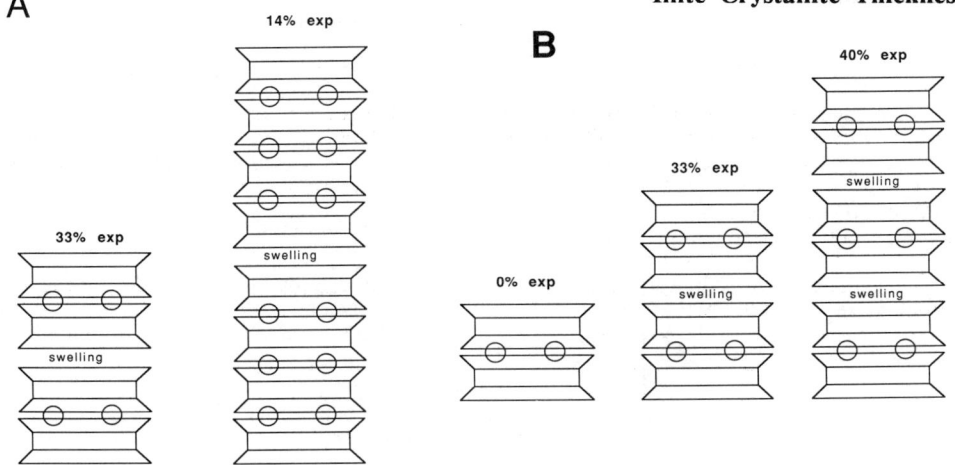

Figure 10. Factors that influence the expandability of MacEwan crystallites: **A**. expandability depends on the thickness of fundamental illite particles in the stacks; **B**. expandability depends on the number of illite particles in the stacks.

The formal relation between XRD expandability (X, in %), mean particle thickness ($\overline{T}$, in nm, for defect-free illite particles), and mean number of illite particles (S) composing MacEwan crystallites is (Eberl and Srodon, 1988):

$$X = \frac{100(S-1)}{(S\overline{T}-1)}. \tag{20}$$

Generally, S varies with expandability, from about 1.4 for I/S of small expandability, to >5 for I/S of large expandability (Eberl and Srodon, 1988). In other words, thinner illite particles often stack better than thicker illite particles. Particle surface roughness and X-Y areas also can influence S. An S less than 2 implies that the sample contains some illite particles that are not articulated in MacEwan crystallites. The presence of defects (Figure 1C) within illite particles that form MacEwan crystallites adds a further complication to Eq. (20).

Figure 11 further demonstrates the complexity of the relation between expandability and particle thickness by comparing XRD expandability data with mean particle thicknesses measured by the fixed cation method for the same I/S sample set. The large expandability change from 100% to 30% is accompanied by a particle thickness increase of only 1 nm. Similarly, a change in particle thickness from 3 to 18 nm is accompanied by an expandability change of less than 5%. Although the expandabilities of illite/smectites change regularly with depth in sedimentary basins (e.g. Perry and Hower, 1970; Hower et al., 1976), and in proximity to igneous intrusions and in regional metamorphic terrains (Nadeau and Reynolds, 1981), expandability probably is not an appropriate measurement to quantitatively determine mass balances or the kinetics of the reaction of smectite to illite.

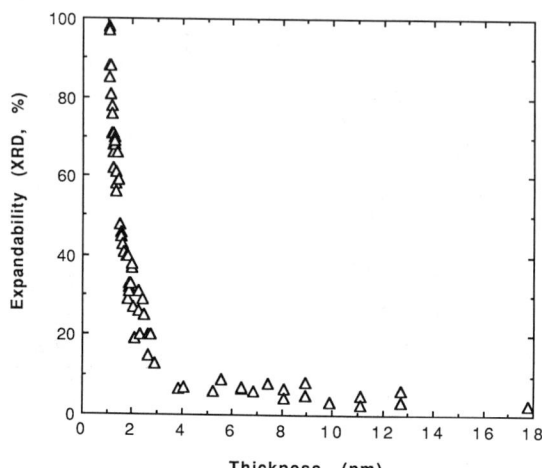

Figure 11. Comparison between particle thicknesses measured by the fixed cation method, and XRD expandabilities (data from Srodon et al., 1986, and Eberl et al., 1987). These data were used to determine Eq. (17) in the text.

The Kubler index, which was defined above as the width of the 001 illite XRD reflection at half height (Kisch, 1983), is subject to all of the factors discussed previously that influence the breadth of XRD peaks. Instrumental broadening can be removed by using a standard (Eq. 2), and the influence of swelling on the 001 reflection can be minimized by using Na-saturated, air-dried samples that are swollen with one water layer (Table 1). Laboratory humidity should range between about 20% and 50% to ensure a one water-layer interlayer complex (MacEwan and Wilson, 1980). The Kubler index may be a rough empirical measure of mean particle thickness, but is subject to the limitations of the Scherrer technique.

SUMMARY AND CONCLUSIONS

The Warren-Averbach method is a convenient and accurate technique for measuring illite X-ray scattering domain thicknesses. This method can be performed using a single XRD peak, the 002 reflection for air-dried, Ca- or Sr-saturated illite samples, because the shape of this peak is not affected by strain or by swelling. Domain thicknesses determined by this method correlate to fundamental illite particle thicknesses for mean thicknesses ranging from about 2 to about 20 nm. Because X-ray scattering domain thickness and fundamental particle thickness may not be equal, it would be prudent to check crystallite thicknesses measured by the W-A technique by another method, such as by SFM or TEM, or by making the appropriate measurements and solving equations 3, 6, 7, 8 or 10 for $\overline{T}$.

ACKNOWLEDGMENTS

We thank R. DeAngelis, L. Heller-Kallai, D. Martin, H. May, R. Reynolds, J. Srodon, S. Tsipursky, J. Walker, and G. Zorn for their reviews of the manuscript, and H. Taylor and J. Srodon for supplying particle size distribution data and photos.

REFERENCES CITED

Altaner, S. P., Weiss, C. A., Jr. and Kirkpatrick, R. J. (1988) Evidence from ^{29}Si NMR for the structure of mixed-layer illite/smectite clay minerals: *Nature* **331**, 699-702.

Baronnet, A. (1982) Ostwald ripening in solution. The case of calcite and mica: *Estudios geologicos* **38**, 185-198.

Barrett, C. S. and Massalski, T. B. (1966) *Structure of Metals*: McGraw-Hill Book Co., New York, 654 pp.

Blum, A. E. and Eberl, D. D. (1992) Determination of clay particle thicknesses using scanning force microscopy: in *Water-Rock Interaction*, Kharaka, T. K. and Maest, A. S., eds., Balkema, Rotterdam, 133-136.

Brindley, G. W. (1980) Order-disorder in clay mineral structures: in *Crystal Structures of Clay Minerals and their X-ray Identification*, Brindley, G. W. and Brown, G., eds., Mineralogical Society, London, 125-195.

Eberl, D. D., Srodon, J., Lee, M., Nadeau, P. and Northrop, H. R. (1987) Sericite from the Silverton caldera, Colorado: correlation among structure, composition, origin, and particle thickness: *American Mineralogist* **72**, 914-934.

Eberl, D. D. and Srodon, J. (1988) Ostwald ripening and interparticle diffraction effects for illite crystals: *American Mineralogist* **73**, 1335-1345.

Eberl, D. D., Srodon, J., Kralik, M., Taylor, B. E. and Peterman, Z. E. (1990) Ostwald ripening of clays and metamorphic minerals: *Science* **248**, 474-477.

Ergun, S. (1970) X-ray scattering from very defective lattices: *Physical Review B.* **131**, 3371-3380.

Gregg, J. M., Howard, S. A. and Mazzullo, S. J. (1992) Early diagenetic recrystallization of Holocene (<3000 years old) peritidal dolomites, Ambergris Cay, Belize: *Sedimentology* **39**, 143-160.

Güven, N. (1972) Electron optical observations on Marblehead illite: *Clays & Clay Minerals* **20**, 83-88.

Hower, J. and Mowatt, T. C. (1966) The mineralogy of illites and mixed-layer illite/montmorillonite: *American Mineralogist* **51**, 825-854.

Hower, J., Eslinger, E. V., Hower, M. and Perry, E. A. (1976) Mechanism of burial metamorphism of argillaceous sediment: 1. Mineralogical and chemical evidence: *Geolical Society America Bulletin* **87**, 725-737.

Hunziker, J. C., Frey, M., Clauer, N., Dallmeyer, R. D., Friedrichsen, H., Flehmig, W., Hochstrasser, K., Roggwiler, P. and Schwander, H. (1986) The evolution of illite to muscovite: mineralogical and isotopic data from the Glarus Alps, Switzerland: *Contributions to Mineralogy and Petrology* **92**, 157-180.

Inoue, A., Velde, B., Meunier, A. and Touchard, G. (1988) Mechanism of illite formation during smectite to illite conversion in a hydrothermal system: *American Mineralogist* **73**, 1325-1334.

Jahren, J. S. (1991) Evidence of Ostwald ripening related recrystallization of diagenetic chlorites from reservoir rocks offshore Norway: *Clay Minerals* **26**, 169-178.

Kisch, H. J. (1983) Mineralogy and petrology of burial diagenesis (burial metamorphism) and incipient metamorphism in clastic rocks: in *Diagenesis in Sediments and Sedimentary Rocks*, Larsen, G. and Chilingar, G. V., eds., Developments in Sedimentology **25B**, Elsevier, New York, 289-494.

Klug, H. P. and Alexander, L. E. (1974) *X-ray Diffraction Procedures for Polycrystalline and Amorphous Materials*, John Wiley and Sons, New York, 966 pp.

Lanson, B. and Champion, D. (1991) The I/S-to-illite reaction in the late stage diagenesis: *American Journal of Science* **291**, 473-506.

Lindgreen, H., Garnaes, J., Hansen, P. L., Besenbacher, F., Laaegsgaard, E., Stensgaard, I., Gould, S. A. C. and Hansma, P. K. (1991) Ultrafine particles of North Sea illite/smectite clay minerals investigated by STM and AFM: *American Mineralogist* **76**, 1218-1222.

MacEwan, D. M. C. and Wilson, M. J. (1980) Interlayer and intercalation complexes of clay minerals: in *Crystal Structures of Clay Minerals and their X-ray Identification*, Brindley, G. W. and Brown, G., eds., Mineralogical Society, London, 197-248.

Méring, J. (1949) L'intérference des rayons X dans les systems à stratification désordonnée: *Acta Crystallographica* **2**, 371-377.

Moore, D. M. and Reynolds, R. C., Jr. (1989) *X-ray Diffraction and the Identification and Analysis of Clay Minerals*: Oxford University Press, New York, 332 pp.

Nadeau, P. H. (1985) The physical dimensions of fundamental clay particles: *Clay Minerals* **20**, 499-514.

Nadeau, P. H. (1987) Relationships between the mean area, volume and thickness for dispersed particles of kaolinites and micaceous clays and their application to surface area and ion exchange properties: *Clay Minerals* **22**, 351-356.

Nadeau, P. H. and Reynolds, R. C. Jr. (1981) Burial and contact metamorphism in the Mancos shale: *Clays & Clay Minerals* **29**, 249-259.

Nadeau, P. H. and Tait, J, M. (1987) Transmission electron microscopy: in *A Handbook of Determinative Methods in Clay Mineralogy*, Wilson, M. J., ed., Chapman and Hall, New York, Chap. 6, 209-247.

Nadeau, P. H., Wilson, M. J., McHardy, W. J. and Tait, J. M. (1984) Interstratified clay as fundamental particles: *Science* **225**, 923-925.

Perry, E. and Hower, J. (1970) Burial diagenesis in Gulf Coast pelitic sediments: *Clays & Clay Minerals* **18**, 165-177.

Reynolds, R. C., Jr. (1985) *NEWMOD©, a Computer Program for the Calculation of One-Dimensional Diffraction Patterns of Mixed-Layered Clays*: R. C. Reynolds, Jr., 8 Brook Dr., Hanover, New Hampshire, 03755.

Siemens (1990) *Diffrac 5000 Powder Diffraction Evaluation Software Reference Manual*, release 2.2, part no. 269-00200, Siemens Analytical Instruments, Inc., 6300 Enterprise Lane, Madison, WI, 53719, p. 16.15.

Srodon, J. (1980) Precise identification of illite/smectite interstratifications by X-ray powder diffraction: *Clays & Clay Minerals* **28**, 401-411.

Srodon, J. (1984) X-ray identification of illitic materials: *Clays & Clay Minerals* **32**, 337-349.

Srodon, J. and Eberl, D. (1984) Illite: in *Micas*, Bailey, S. W., ed., *Reviews in Mineralogy* **13** Mineralogical Society of America, Washington, D.C., 495-544.

Srodon, J., Morgan, D. J., Eslinger, E. V., Eberl, D. D. and Karlinger, M. R. (1986) Chemistry of illite/smectite and end-member illite: *Clays & Clay Minerals* **34**, 368-378.

Srodon, J., Andreoli, C., Elsass, F. and Robert, M. (1990) Direct high-resolution electron microscopic measurements of expandability of mixed-layer illite/smectite in bentonite rock: *Clays & Clay Minerals* **38**, 373-379.

Srodon, J., Elsass, F., McHardy, W. J. and Morgan, D. J. (1992) Chemistry of illite/smectite inferred from TEM measurements of fundamental particles: *Clay Minerals* **27**, 137-158.

Stokes, A. R. (1948) A numerical Fourier-analysis method for the correction of widths and shapes of lines on X-ray powder photographs: *Proceedings of the Physical Society (London)* **61**, 382-391.

Warren, B. E. and Averbach, B. L. (1950) The effect of cold-work distortion on X-ray patterns: *Journal of Applied Physics* **21**, 595-599.

Warren, B. E. and Averbach, B. L. (1953) The diffuse scattering of X-rays: in *Modern Research Techniques in Physical Metallurgy*, American Society for Metals, Cleveland, 95-130.

Warren, B. E. (1959) X-ray studies of deformed metals: *Progress in Metal Physics* **8**, Pergamon Press, London, 147-202.

Warren, B. E. (1969) *X-ray Diffraction*, Addison-Wesley, London, 381 pp.

Eberl and Blum

APPENDIX: THE WARREN-AVERBACH METHOD

The Warren-Averbach method for determining crystallite size and strain has been derived by Warren and Averbach (1950), Warren (1969), and Klug and Alexander (1974). This appendix describes how the method works, but does not give a rigorous derivation of the theory. Much of the following discussion follows that of Warren (1953, 1969), Barrett and Massalski (1966), and an unpublished, anonymously authored (by G. Zorn), Siemens Corporation document entitled "Chapter 16, Crystallite Size/Microstress."

The steps involved in the W-A analysis are as follows: 1) A series of XRD peaks related by n in Bragg's law are chosen for analysis in order to find the scattering domain size for this dimension in the crystallites. Here we are concerned with 00ℓ reflections for measuring illite crystallite thicknesses. 2) These diffraction peaks are smoothed and modeled using, for example, a split Pearson VII function. During this step, the $K\alpha 2$ peaks are removed, and the backgrounds and shapes of the $K\alpha 1$ peaks are determined. 3) Peak shapes are corrected for theta-dependent functions such as the Lorentz-polarization factor and the structure factor. 4) Peak shapes are corrected for machine broadening by use of a standard that has been analyzed and modeled in exactly the same manner as the sample. 5) The corrected profile for each reflection is modeled using a fast Fourier transform. 6) The Fourier coefficients found in step 5 are corrected to remove the strain broadening contribution. During this calculation, the root mean square of the strain and the presence of stacking faults can be determined. 7) The corrected Fourier coefficients are used to determine the mean crystallite size and the crystallite size distribution for a sample. Steps 2 through 7 above are applied automatically by using the Siemens Corporation software programs FIT and CRYSIZ.

The W-A method relies on the fact that the distribution of diffracted X-ray photons in a Debye-Scherrer cone ($P'_{2\theta}$) can be expressed as a Fourier series. The following discussion uses the mathematical notation of Warren, and identifies Warren's "cells" with one nm thick illite layers, and his "columns" with illite crystallites that diffract coherently perpendicular to the layers. The expression for the distribution of diffracted X-rays is:

$$P'_{2\theta} = K(\theta)N \sum_{n=-\infty}^{+\infty} \left(A_n \cos 2\pi n h_3 + B_n \sin 2\pi n h_3\right), \qquad (21)$$

where $K(\theta)$ is a complicated term that includes the structure factor, the Lorentz-polarization factor, atomic scattering factors, multiplicity factors, diffraction geometry, etc. (Barrett and Massalski, 1966; Warren, 1969). $K(\theta)$ is removed from the experimental peak in step 3 above, and therefore the Fourier coefficients A_n and B_n are concerned with corrected intensities. N is the total number of diffracting layers in the sample, a constant. The other terms in Eq. (21) are defined as follows: $h_3 = \dfrac{2|\mathbf{a}_3|\sin\theta}{\lambda}$, where λ is the wavelength of the radiation, and $\mathbf{a}_3$ is the vector of the unit cell perpendicular to the layers. $\mathbf{a}_3$ is determined experimentally from the Warren formula $\dfrac{2a_3}{\lambda}\left(\sin\theta_{max} - \sin\theta_{min}\right) = 1/2$, where $\sin\theta_{max}$ = the Bragg angle at maximum peak intensity, and $\sin\theta_{min}$ = the Bragg angle at background near the peak. n is the harmonic number corresponding to the number of layers in a particle, and is related to the structure by $L = na_3$, where L is the particle thickness normal to the diffracting planes in the crystal.

The Fourier coefficients A_n and B_n in Eq. (21) each are the product of two terms:

$$A_n = \left(\frac{N_n}{N_3}\right) \cdot \langle \cos 2\pi \ell Z_n \rangle, \text{ and} \qquad (22)$$

$$B_n = -\left(\frac{N_n}{N_3}\right) \cdot \langle \sin 2\pi \ell Z_n \rangle. \qquad (23)$$

N_3 is the average number of layers per crystallite in the sample (the quantity we want to find), and equals $N/N_{(col)}$, where $N_{(col)}$ is the total number of illite crystallites in the sample. N_n is the average number of layer pairs per particle = N_n(sample)/$N_{(col)}$, where N_n(sample) = number of layers in the whole sample in particles having an nth neighbor in the same column. ℓ is the order of the observed reflection. Z_n is the relative distortion (strain) between layers that are n layers apart, measured in units of the cell axis perpendicular to the layers.

If there are equal probabilities for positive and negative Z_n values in the sample, then the sine term in Eq. (23) will be zero, the XRD peaks will be symmetric (assuming that there are no stacking faults), and Eq. (23) can be ignored. The Fourier coefficient (A_n) in Eq. (22) is taken to be the product of a particle size coefficient:

$$A_n^S = \frac{N_n}{N_3}, \qquad (24)$$

and a distortion coefficient:

$$A_n^D = \langle \cos 2\pi \ell Z_n \rangle, \qquad (25)$$

where the brackets refer to the average value. The size coefficient (Eq. 24) is not a function of the order of the reflection, whereas the distortion coefficient (Eq. 25) is a function of ℓ, and becomes unity for the zeroth order reflection. Therefore, measurement of a series of 00ℓ reflections, and extrapolation of the Fourier coefficients for each particle thickness to the zeroth order removes the strain contribution from the Fourier coefficients.

This calculation is performed as follows. Combining Eqs. (22), (24) and (25), and converting to natural logarithms gives:

$$\ln A_n(\ell) = \ln A_n^S + \ln A_n^D(\ell). \qquad (26)$$

Combining Eqs. (25) and (26), and expanding the strain term (which is allowed for small values of ℓ and Z_n) yields:

$$\ln A_n(\ell) = \ln A_n^S - 2\pi^2 \langle Z_n^2 \rangle \cdot \ell^2. \qquad (27)$$

For a fixed value of n (the number of layers in a particle), we plot $\ln A_n(\ell)$ (determined during the Fourier analysis) as a function of ℓ^2 (Figure A1). For small values of ℓ, the curves will be straight lines for which the intercepts at $\ell^2 = 0$ give the values of the size coefficients, $\ln A_n^S$, and for which the slopes give $-2\pi^2 \langle Z_n^2 \rangle$, from which the strains can be determined. Because $\langle Z_n^2 \rangle$ is expressed in terms of units of the cell axis, and because the strain has both positive and negative values, strain is rather expressed as the root mean square of the microstrain = $\sqrt{\langle \varepsilon_L^2 \rangle}$, where $\varepsilon_L = \dfrac{a_3 Z_n}{a_3 n}$, where $a_3 n$ is the undistorted particle thickness that is changed by distortion by a distance of $a_3 Z_n$. The brackets refer to the average value. The Siemens program CRYSIZ prints out and plots $\sqrt{\langle \varepsilon_L^2 \rangle}$ as a function of distance from the point that causes the strain (e.g. a dislocation).

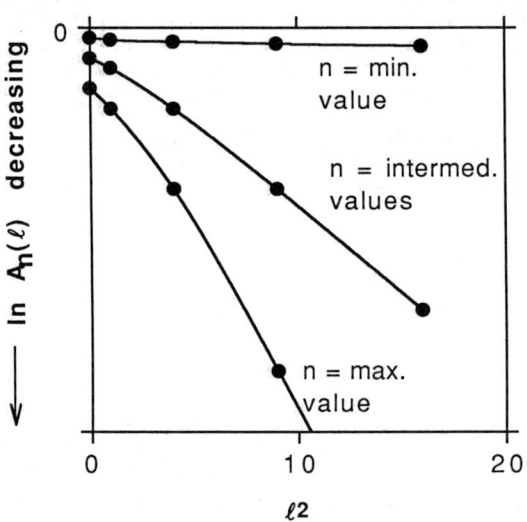

Figure A1. Hypothetical plot of Fourier coefficients (ln $A_n(l)$) as a function of the square of the order of the reflection (l^2). Extrapolation to $l^2 = 0$ gives the corrected Fourier coefficients (with the strain component removed) as a function of the number of layers in the crystallite (n).

If XRD peak broadening was entirely due to strain broadening, then all of the curves in Figure A1 would intersect the origin at $l^2 = 0$. Conversely, if all of the broadening was related to particle size broadening, then the curves would be horizontal lines. The departure of points from smooth lines in Figure A1 is a measure of stacking fault broadening for particular reflection orders. For illite, strain broadening was shown to be insignificant, and a peak was chosen for analysis so as to minimize stacking fault broadening due to swelling. Therefore, Fourier analysis of a single peak (002 peak for air-dried, Ca-saturated sample) is sufficient to determine $\ln A_n^S$ for illites. Eberl and Srodon (1988) mistakenly used the 002 and 005 for glycol-solvated samples for W-A analysis, and confused broadening related to swelling with that caused by strain.

To continue the analysis of particle thickness, $\ln A_n^S$ values determined from Figure A1 are plotted as a function of particle thickness, as depicted in Figure A2. Eq. (24) can be rewritten:

Illite Crystallite Thickness

$$A_n^S = \frac{1}{N_3} \int_{i=|n|}^{\infty} (i-|n|)p(i)di,\qquad(28)$$

where p(i) is the fraction of particles having a thickness of i layers. The first derivative of Eq. (28) is:

$$\frac{dA_n^S}{dn} = -\frac{1}{N_3} \int_{i=|n|}^{\infty} p(i)di.\qquad(29)$$

For $|n| = 0$, the integral over the distribution function p(i) is equal to unity, and therefore N_3 can be determined from the initial slope of a plot of A_n^S versus n. In practice, n is replaced by L, the particle thickness, where $L = a_3 n$. A_n^S thereby becomes A_L^S in the plot in Figure A2.

In the plot of A_L^S versus L, the first coefficients in the curve usually are too small, because information was lost from the tails of the XRD peaks for very small n. This hook effect (Figure A2) is corrected by a linear extrapolation of the steepest part of the curve to L = 0 and normalizing the ordinate to $A_L^S = 1.0$ at L = 0. The intersection of this line with the abscissa gives the mean particle thickness.

The second derivative of A_n^S with respect to n (Eq. 28) yields:

$$\left(\frac{d^2 A_n^S}{dn^2}\right) = \frac{p(|n|)}{N_3},\qquad(30)$$

which is the crystallite size distribution for illite, given by the slope of the curve in Figure A2. CRYSIZ prints both the relative frequency distribution and the cumulative distribution.

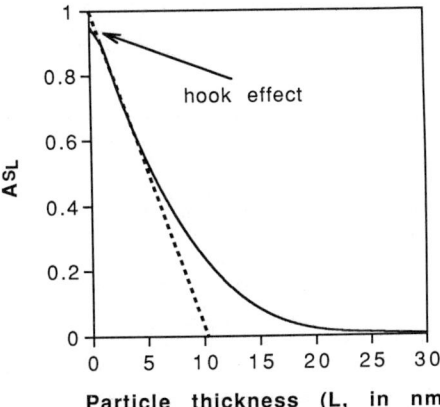

Figure A2. Hypothetical plot of A_L^S versus particle thickness ($L = a_3 n$). Extrapolation (dashed line) of the steepest part of the curve to L = 0 defines the ordinate at 1.0. Then extrapolation to $A_L^S = 0$ gives the mean particle thickness, about 10 nm. The derivative of the combined curve passing through $A_L^S = 1.0$ and tailing off to L > 30 gives the particle size distribution.

A COMPUTER TECHNIQUE FOR RAPID DECOMPOSITION OF X-RAY DIFFRACTION INSTRUMENTAL ABERRATIONS FROM MINERAL LINE PROFILES

R. C. Jones and H. U. Malik

CONTENTS

Introduction	156
Symmetrical And Split Pseudo-Voigt Functions	157
Method Of Minimization	159
Methods And Materials	161
Curve Fitting Examples	161
Conclusions	169
References Cited	170

A COMPUTER TECHNIQUE FOR RAPID DECOMPOSITION OF X-RAY DIFFRACTION INSTRUMENTAL ABERRATIONS FROM MINERAL LINE PROFILES

R. C. Jones and H. U. Malik

Department of Agronomy and Soil Science
University of Hawaii at Manoa
G. Donald Sherman Laboratory
1910 East-West Road
Honolulu, Hawaii 96822

INTRODUCTION

The convolution of instrumental aberrations with the true mineral line profile is represented by $P(2\theta) = [(W * G) * S](2\theta)$ + background (Balzar, 1992; Huang and Parrish, 1975). W is the wavelength distribution function, G is all other instrumental functions, S is the specimen profile that is caused by pure physical broadening, $*$ represents the convolution operations, and the convolution operation is over the 2θ scale. For instrumental effects, Alexander (1954) calculated a convolution total of six instrumental line profiles that are reproduced by Klug and Alexander (1974, page 292). They showed that there was excellent agreement between a theoretical profile and that of the 101 quartz profile. The five instrumental profiles that contribute to the observed line profile are: the source (represented by W above) which is approximated by a *symmetrical* Gaussian distribution; flat specimen surface which is an *asymmetrically* truncated function that affects the lower 2θ side of a peak; axial divergence which is *asymmetrical* and also affects the lower 2θ side of a peak; specimen transparency which is also *asymmetrical* and affects the lower 2θ side of a peak; and the effects of the receiving slit and misalignment which are *symmetrical* distributions. These five instrumental aberrations are represented by G above. Inasmuch as flat specimen effects, axial divergence, and specimen transparency contribute to the asymmetric character of diffraction profiles, we attribute all asymmetry on the lower 2θ side of the profiles used as examples in this chapter as due to instrumental aberrations. Such an assumption is certainly not valid for all minerals. The materials selected as examples for this chapter are believed to have symmetrical mineral profiles.

Computer program *Pi'o Pili Pa'a* [Hawaiian for "curve close fit"] (Jones, 1989) performs a decomposition of $(W * G)$ from S by minimizing the error between an experimentally determined profile and the summation of two or more profiles representing instrumental aberrations and the pure specimen profile. The profile function and background options available are listed in Tables 1 and 2 respectively.

For the decomposed patterns presented in this chapter, peaks representing instrumental aberrations were refined as one or two split pseudo-Voigt distributions. In order to account for asymmetry, peaks below 30 °2θ were fit with two split pseudo-Voigt peaks and those above were fit with one split pseudo-Voigt peak because of less asymmetry. The peak representing the mineral contribution to the experimental profile was fit with one symmetrical pseudo-Voigt distribution. Therefore, only the descriptions of the pseudo-Voigt function will be given here.

Table 1. The distributions that can be fit by computer program *Pi'o Pili Pa'a*.

Symmetrical pseudo-Voigt
Pure Gaussian with no Ka2 component
Pure Cauchy with no Ka2 component
Split pseudo-Voigt
Symmetrical Pearson type-VII
Split Pearson type-VII
Pearson type-VII/pseudo-Voigt
Pseudo-Voigt/modified exponential

Table 2. The background options that can be refined by *Pi'o Pili Pa'a*

Lorentz-polarization factor
Straight, zero-slope line background
Straight, sloping line background
Quadratic distribution
Gaussian distribution
Cauchy distribution
Gaussian/modified exponential distribution

SYMMETRICAL AND SPLIT PSEUDO-VOIGT FUNCTIONS

Figure 1 is a synthetically produced profile that illustrates the differences between Gaussian and Cauchy distributions. The peak in Figure 1 is described by the following relationships from Langford et al. (1986), Henke et al. (1983), and Fraser and Suzuki (1969, 1970). The Gaussian distribution is expressed as

$$I_G = I_0 \exp(-ln2(2\Delta2\theta)/\Gamma)^2,$$

and the Cauchy distribution is expressed as

$$I_C = I_0/(1 + ((2\Delta2\theta)/\Gamma)^2),$$

where I_0 is the intensity of the Kα1 peak, $\Delta2\theta$ is the step increment on the 2θ scale, and Γ is the full width at half maximum (FWHM) of the Kα1 component of the peak.

Wertheim et al. (1974) proposed the pseudo-Voigt distribution as a linear combination of Gaussian and Cauchy distributions in lieu of a rigorous convolution of the two functions. Although the pseudo-Voigt function was originally proposed for the analysis of Mössbauer spectra, it has been very successfully employed for X-ray spectra by Langford et al. (1986),

Henke et al. (1983), and Young and Wiles (1982) among others. The Gaussian and Cauchy distributions are linearly combined to form the pseudo-Voigt distribution by

$$I_V = I_G \delta + I_C(1 - \delta),$$

where I_V is the pseudo-Voigt intensity distribution and δ is the Gaussian-Cauchy mixing parameter. The mixing parameter, δ, is allowed to vary from 0.0 for a pure Cauchy distribution to 1.0 for a pure Gaussian distribution. The pseudo-Voigt combination of Gaussian and Cauchy distributions is valid only if the peak widths are identical.

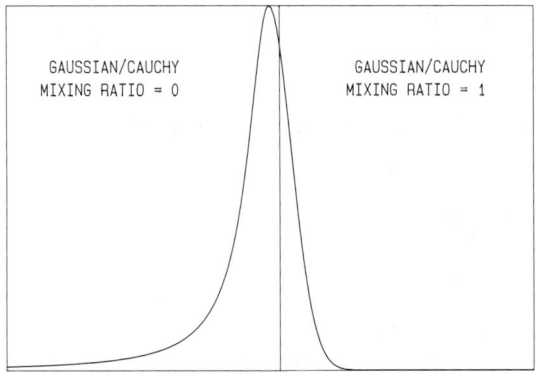

Figure 1. Synthetically generated instrumental peak profile having a pure Cauchy distribution on the left side and a pure Gaussian distribution on the right side. The vertical line is the position of a mineral peak if one were present.

Figure 1 is a split pseudo-Voigt distribution that was produced at the two limits of the Gaussian-Cauchy mixing parameter, δ. Figure 1 is, therefore, a simulated instrumental profile having a Cauchy distribution ($\delta = 0.0$) on the left side and a Gaussian distribution ($\delta = 1.0$) on the right side. The straight vertical line near the center of the peak is the exact position of a mineral peak if one were represented in the figure. Klug and Alexander (1974, p. 292) show a convolved instrumental profile shifted more to the left of a theoretical mineral peak position than is shown in Figure 1. Experience with the authors' diffractometer has shown that a peak representing instrumental effects typically resides to the left (lower 2θ side) of the mineral peak by no more than about 0.05 °2θ. In Figure 1 the instrumental peak is displaced 0.04 °2θ to the left of the mineral peak position (vertical line).

Figure 2 is a simulated pure mineral profile. The peak is symmetrical and has a Gaussian/Cauchy mixing ratio, δ, of 0.5 (one-half Gaussian and one-half Cauchy). In the case of Figure 2, the vertical line representing the true position is at the exact center of the peak. Peak profiles produced by diffractometers are a convolution of one or more peaks that represent instrumental aberrations and the true mineral peak profile. By curve fitting with computer program

Rapid Decomposition of Instrumental Aberrations

Pi'o Pili Pa'a, profiles approximating instrumental effects and mineral peaks may be decomposed, the results of which are listings of the parameters that describe the various profiles.

METHOD OF MINIMIZATION

Minimization of errors by computer program *Pi'o Pili Pa'a* is accomplished in the following manner. First, R is calculated by

$$R = \sum_{i=1}^{n} ((I_i - I_c)/\omega_i)^2,$$

where I_i is the intensity of the experimental data at step i, I_c is the calculated intensity at step i, and ω_i is the weight that is applied to the minimization. Three different weighting factors are employed by *Pi'o Pili Pa'a* for minimization: 1) when the weight (value of ω_i) is 1.0 for all intensities, R is the error sum of squares and all intensities are treated with equal emphasis; 2) when $\omega_i = I_i^{1/2}$ then $\omega_i = s_i$ (standard deviation of the intensities) and R is the chi-square, in which case the greatest emphasis is placed on the peak tails and background; and 3) when $\omega_i = 1/I_i^{1/2}$ then ω_i is the inverse of s_i, R is then the inverse chi-square and the greatest emphasis is placed on the peak apices. RSTAT is then the reduced statistic that is found by

$$\text{RSTAT} = R/(n - N_{\text{var}}),$$

where n is the number of data pairs and N_{var} is the number of parameters that are allowed to vary during the minimization. Those parameters that are held constant by the program user are not included in N_{var}.

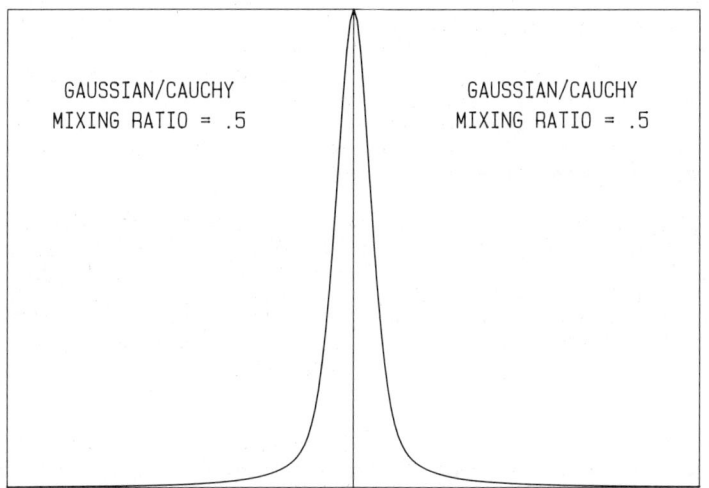

DEGREES 2-THETA

Figure 2. Synthetically generated mineral peak having a 50:50 mixture of Cauchy and Gaussian distributions on both sides.

Jones and Malik

R is minimized during the fitting process using the method of Marquardt (1963; see also Bevington, 1969, p. 232-242). The details of the Marquardt algorithm are explained here in the context used in *Pi'o Pili Pa'a*. First, the initial error, R_1 is found; then, deviating from the Marquardt algorithm, all of the fitting parameters are normalized to values between 0.0 and 1.0 by

$$Q_j = (P_j - P\text{min}_j)/(P\text{max}_j - P\text{min}_j),$$

where Q_j is the normalized parameter array, P_j is the non-normalized parameter array, $P\text{max}_j$ is the maximum constraint that is placed on each variable parameter by the program user, $P\text{min}_j$ is the minimum constraint, and the subscript, j, represents all of the variable parameters. For example, an intensity of 5,000 counts is given the same weight as a peak width parameter of 0.1 °2θ after normalization. Next, an array of gradient vectors and a curvature matrix are generated (see Bevington, 1969, for a complete explanation). Then, a constant factor, $f\lambda$ (Bevington, 1969), is added to the diagonal terms of the curvature matrix. When $f\lambda$ is small, inversion of the curvature matrix will approach an analytical solution. Marquardt recommends a starting value of 0.001 for $f\lambda$. With the inversion of the curvature matrix a new parameter array is generated which is denormalized by

$$P_k = Q_k \bullet (P\text{max}_j - P\text{min}_j)/(P\text{max}_j - P\text{min}_j),$$

where P_k is a new non-normalized parameter array and Q_k is a new normalized parameter array. $P\text{max}_j$ and $P\text{min}_j$ are the unchanged maximum and minimum constraints that were originally set by the program user. The error after curvature matrix inversion, R_2, is then computed using the new non-normalized parameters. If $R_2 > R_1$, $f\lambda$ is increased by a factor of 10, and added to the diagonal terms of the curvature matrix which is again inverted producing another parameter array. If R_2 is still greater than R_1, $f\lambda$ is again increased by a factor of 10 and the curvature matrix is reinverted. The value of $f\lambda$ is incremented by a factor of 10 until either $R_2 < R_1$ or $f\lambda$ reaches a value 10^{10} at which time the fitting process will stop and a "FITTING FAILURE" message will appear on the screen. As the value of $f\lambda$ increases the algorithm approaches a gradient-search method with a diminishing step size. Therefore, if the value of $f\lambda$ becomes too large the theory of the algorithm predicts that a minimum may not be reached. If $R_2 < R_1$, $f\lambda$ is decreased by a factor of 10, and the new parameters array is the starting point for the next iteration. By decreasing the value of $f\lambda$, the step size toward a minimum is increased.

Convergence occurs when, in the final iteration, $R_2 = R_1$ to the eighth significant digit (the curvature matrix and matrix inversion are computed in double precision). R_1 is always computed at the beginning of each iteration. Therefore, if $R_2 = R_1$ there is no statistical improvement in the fit for the iteration where convergence occurs. A banner appears on the screen announcing successive curvature matrix inversions and the value of $f\lambda$ for each inversion. The program is written so that each time the fitting routine is entered (another iteration) the value of $f\lambda$ is checked. If the value is greater than 1, $f\lambda$ is divided by 1000 before the first curvature matrix inversion cycle to increase the step size to the minimum.

The most common cause of fitting failure is the reversal of a parameter's constraints. For example, if a peak has a nominal intensity of 1,000 counts and the maximum allowable value is inadvertently entered as 500 counts and the minimum allowable value is entered as 1,500 counts, fitting will fail. In such a case R_2 is always larger than R_1. When fitting powder patterns with only a few peaks, errors in assigning constraints are readily apparent. For powder patterns having

Rapid Decomposition of Instrumental Aberrations

up to 20 peaks, tracking down errors in the constraints is difficult. A powder pattern having 20 peaks has up to 212 parameter constraints to check.

METHODS AND MATERIALS

The materials used for the examples in this chapter (Table 3) are either National Institute of Standards and Technology (NIST) Standard Reference Materials (SRM) or materials that we believe contain a minimum of structural disorder. Furthermore, with the exception of the Hawaiian bauxite, the materials used for the illustrations were chosen because they showed no evidence of peak broadening due to small diffracting domain size.

Table 3. The materials used

SRM-675 fluorophlogopite, $d(001) = 9.98$ Å
SRM-676 alumina (corundum structure)
CMS Source Clay CCa-1, ripidolite, Flagstaff Hill
El Dorado County, CA (Post and Plummer, 1972)
magnetite-hematite mixture, Ward's Natural Science Establishment
Hawaiian bauxite (Kapaa soil series, Anionic Acrudox, clayey, gibbsitic, isohyperthermic), Island of Kauai

---

The samples were dry ground to pass a 325-mesh sieve and packed into Philips type cavity mounts (15 x 20 mm). X-ray diffraction patterns were produced by a Philips, vertical axis diffractometer equipped with a theta-compensating divergence slit, an incident beam Soller collimator, a diffracted beam curved pyrolytic-graphite monochromer, and a 2.2 kW long fine-focus copper X-ray tube operated at 40 kV and 45 mA (except as noted), or on a synchrotron powder diffractometer at the National Synchrotron Light Source, Brookhaven National Laboratory. Three X-ray instrumental configurations were used. 1) The samples were analyzed using the theta-compensating divergence slit, a receiving slit size of 0.2 mm, and a Soller collimator was placed in the incident beam only. 2) The samples were analyzed as before with the theta-compensating divergence slit, but with a receiving slit of 0.1 mm and a second Soller collimator in the diffracted beam. 3) The divergence slit was fixed so that at 10 °2θ a sample length of 13 mm was exposed to the X-ray beam. This setting was equivalent to a divergence slit opening of 0.5°. The small divergence slit aperture was used to minimize flat specimen effects. For this configuration the receiving slit and the diffracted beam Soller collimator remained the same as in configuration 2.

All diffraction patterns except those for the Hawaiian bauxite, were collected with a step size of 0.005 °2θ. The Hawaiian bauxite pattern was collected with a step size of 0.025 °2θ by laboratory diffractometer and 0.01 °2θ by synchrotron powder diffraction. Integration times varied with the diffractometer configuration and type of sample.

CURVE FITTING EXAMPLES

The ripidolite 001 pattern Figure 3 was collected with diffractometer configuration 2. The profile is fit with three peaks of which the first two are taken as accounting for instrumental aberrations. The main, right-most peak, therefore, represents the mineral peak whose breadth is due to the physical properties of the mineral. The full width at half maximum (FWHM) of the mineral peak is 0.0781 °2θ (all FWHM values were computed in double precision and apply only to the Kα1 line for all peaks, including the instrumental peaks.) The FWHM of the first instrumental peak is 0.3514 °2θ, and the FWHM of the second instrumental peak is 0.1384 °2θ. A conclusion that can be drawn from the result of curve fitting the ripidolite 001 peak is that a major portion of the breadth, and all of the asymmetry, of the ripidolite peak is due to instrumental aberrations. We found that for peaks below about 20 °2θ there was very little difference in the profiles collected with and without the theta-compensating slit. The same specimen surface length was illuminated by the X-ray beam in both cases (~13 mm) and, therefore, the effects of a flat specimen are essentially the same.

Figures 4 and 5 show the fluorophlogopite 001 peak collected under two different diffractometer configurations. The data for Figure 4 was collected with diffractometer configuration 1. The first instrumental peak has a FWHM of 0.4037 °2θ, and the second instrumental has a FWHM of 0.1340 °2θ. The mineral peak has a FWHM of 0.0841 °2θ. The data for Figure 5 was collected with diffractometer configuration 2. In Figure 5 the first instrumental peak has a FWHM of 0.3241 °2θ, and the second instrumental peak has a FWHM of 0.1079 °2θ. The FWHM of the mineral peak is 0.0788 °2θ.

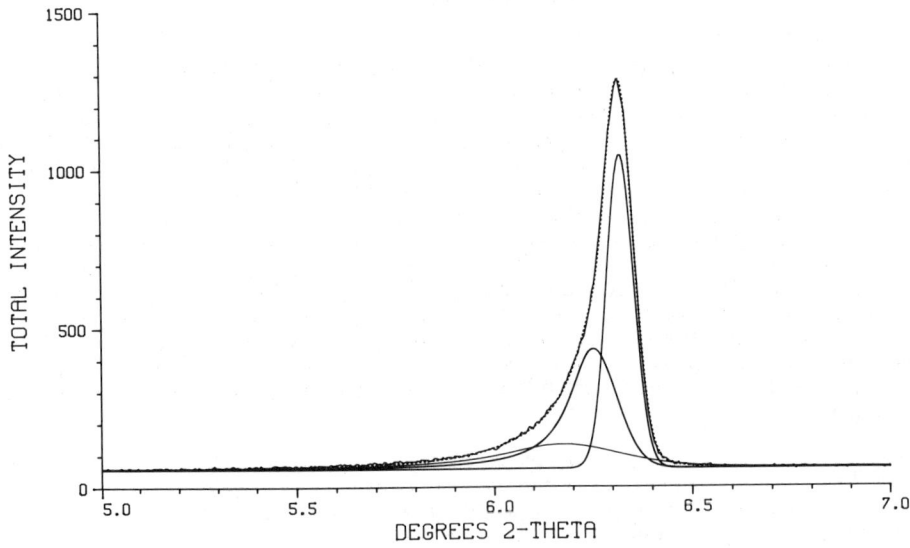

Figure 3. Ripidolite 001 peak fit with two broad, low intensity "instrumental" peaks and a mineral peak. The breadth and shape of the mineral peak was determined by a best fit of a single symmetrical peak to the experimental profile after the two "instrumental" peaks were added.

Rapid Decomposition of Instrumental Aberrations

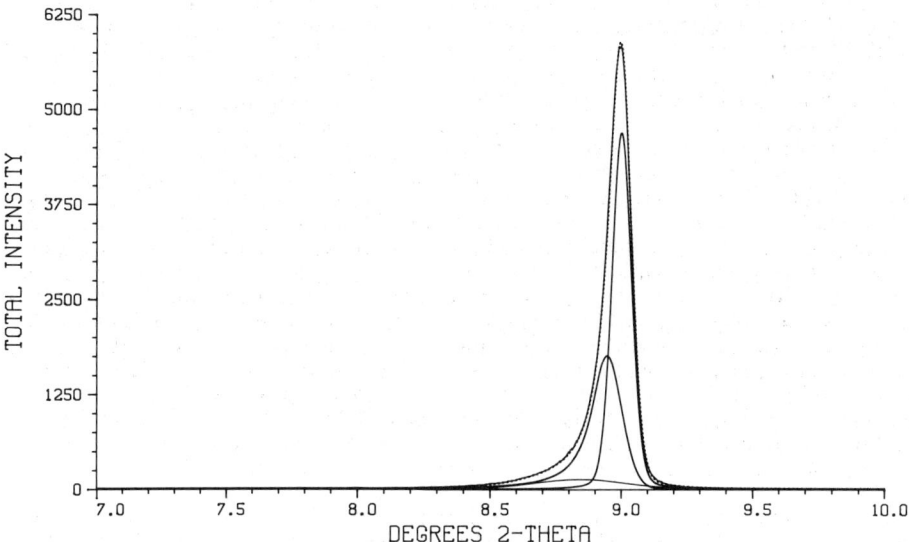

Figure 4. Fluorophlogopite 001 peak collected using diffractometer configuration 1 (see text).

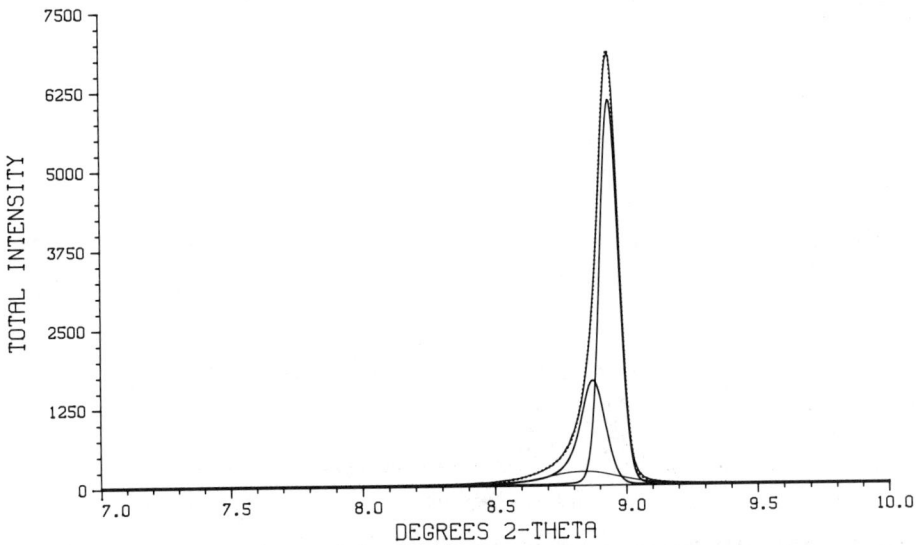

Figure 5. Fluorophlogopite 001 peak collected using diffractometer configuration 2 (see text).

With diffractometer configuration 2, higher resolution data is collected than with configuration 1 and the peaks that represent instrumental aberrations are narrower in Figure 5 than in Figure 4, as expected. If the mineral peak breadth is also narrower with a higher resolution diffractometer configuration, then the breadth of the mineral peak as found by curve fitting is not unique. Its breadth is dependent upon instrumental resolution just as are the peaks that represent instrumental aberrations. When curve fitting a peak representing the physical conditions of a mineral, one might assign a fixed set of parameters for the peak. Then, the instrumental peaks might account for the remainder of the experimental profile. If such a procedure were to be followed, the question that is not answered is what is the true breadth and shape of the mineral peak? Ergun (1968) used the diamond 111 line as a standard for deconvolving instrumental aberrations from a carbon black pattern. He assumed that the true mineral peak had negligible breadth and the observed breadth of the diamond 111 peak could be attributed to instrumental aberrations only. Ergun's premise is based on the assumption that if a substance has close to perfect crystallinity, that is, negligible diffracting domain size broadening and negligible lattice strains, then as in the case of diamond the breadth of a mineral peak is negligible and, therefore, can be ignored. Langford (1968) notes, however, that experimental conditions must be taken into consideration (the resolution of the diffractometer configuration used for the analysis) when interpreting line broadening in powder diffractometry. He cautioned that, when examining polycrystalline specimens, size and lattice strains must also be taken into account. That is to say, most of the materials routinely analyzed by diffractometry do exhibit diffracting domain size broadening and lattice strains.

In Figure 6 the alumina 104 peak is fit with two peaks, the first of which represents instrumental aberrations and the second is taken to be the mineral profile. The fitting philosophy taken is that the fewest number of peaks that will fit the profile adequately (a low error statistic and a good graphical fit) should be used. Any number of peaks, greater than the minimum, can be used for an increasingly closer fit of a profile, however, there may be no physical significance to the additional peaks.

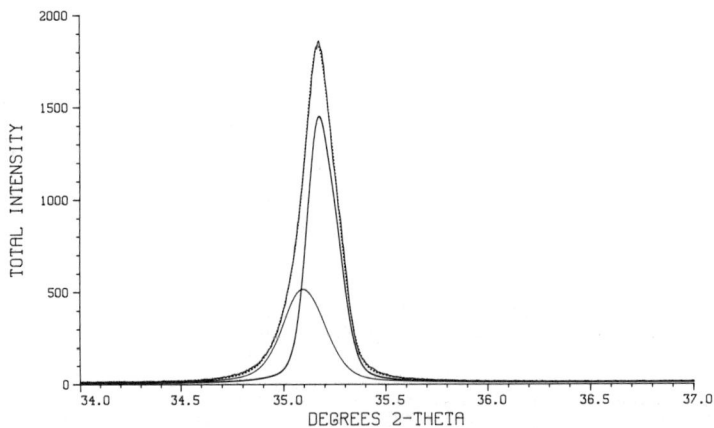

Figure 6. Alumina 104 peak collected using diffractometer configuration 1 (see text).

Rapid Decomposition of Instrumental Aberrations

Diffractometer configuration 1 was used to collect the data for Figure 6 and, as such, two instrumental factors should have been prominent--flat specimen effects and specimen transparency. The FWHM of the instrumental peak is 0.2165 °2θ. The main, or what might be recognized as the mineral peak, has a FWHM of 0.1323 °2θ. In Figure 7 the alumina 104 peak was recorded with diffractometer configuration 3. With this configuration, flat specimen effects should have been minimized. The specimen length exposed to the X-ray beam for the pattern in Figure 6 was 13 mm vs. about 4 mm for the specimen length for the pattern in Figure 7. The FWHM for the instrumental peak in Figure 7 is 0.1280 °2θ. The peak that might be recognized as the mineral peak has a FWHM of 0.0810 °2θ. Because of the angular position of the alumina 104 peak and the narrow breadth of the peak, there is a clear separation of the Kα1α2 doublet.

Inasmuch as the divergence slit was fixed at 0.5° for the pattern in Figure 7 the intensity was greatly reduced from that of the pattern in Figure 6. Intensities for the pattern in Figure 6 were collected with an integration time of 2 seconds per step, whereas the intensities in Figure 7 were collected with an integration time of 10 seconds per step.

Quite clearly, what we are recognizing to be the mineral peak also contains instrumental aberrations as well as the true physical mineral contributions to the peak breadths. This problem is illustrated in Figures 8 and 9. Because of the slowly decaying tails in Figure 8, one symmetrical Cauchy distribution with a FWHM of 0.1 °2θ was tried as a first approximation fit to a hematite 024 peak collected with configuration 2. In this pattern, specimen transparency should not have had a large effect on the shape of the peak because of the high mass absorption of the specimen. There is more than the influence of instrumental aberrations on the hematite 024 peak that is

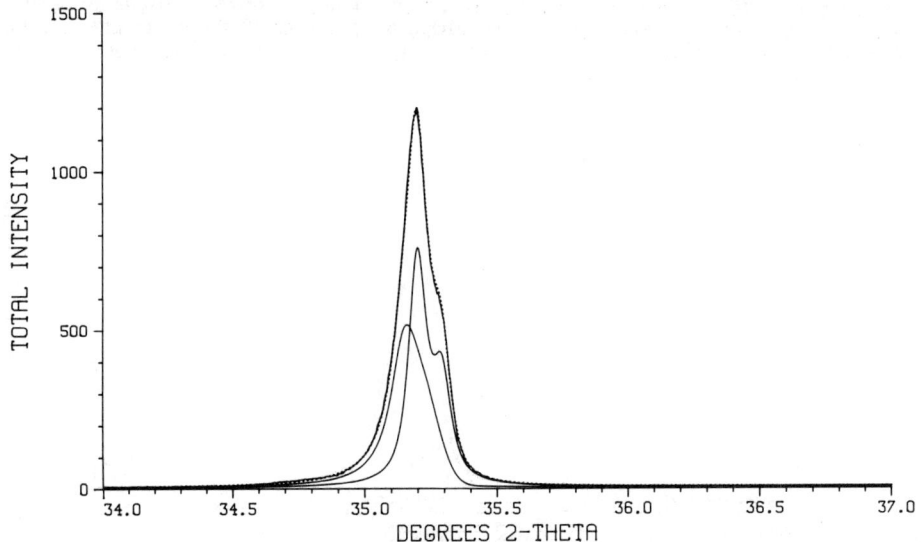

Figure 7. Alumina 104 peak collect using diffractometer configuration 3 (see text). With this configuration, over 50% of the experimental breadth is accounted for by "instrumental" aberrations. Note the difference between the breadths of the mineral peaks between this figure and Figure 6.

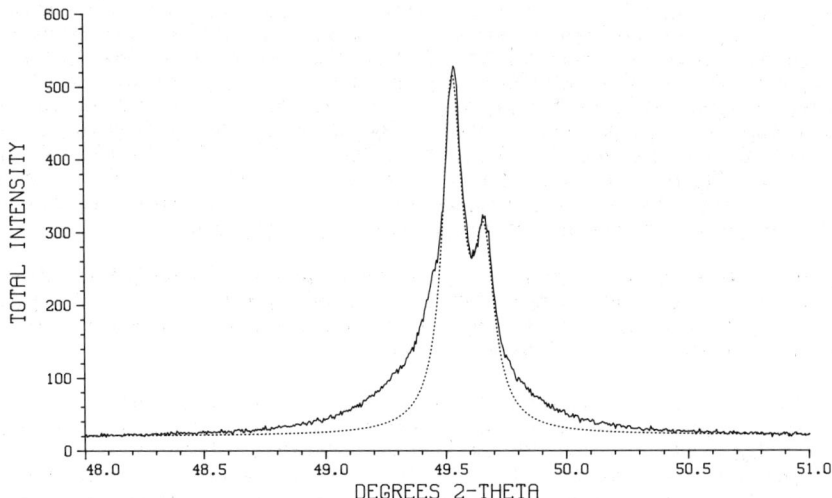

Figure 8. Hematite 024 peak collected using diffractometer configuration 1 (see text). The dotted curve is a best fit of a symmetrical, pure Cauchy distribution. In the previous examples a symmetrical, pseudo-Voigt distribution fit the shape of the upper 2θ side of the peak. In this example there is clear evidence that there is crystal lattice disorder in the mineral because of the misfit in the upper 2θ side of the peak.

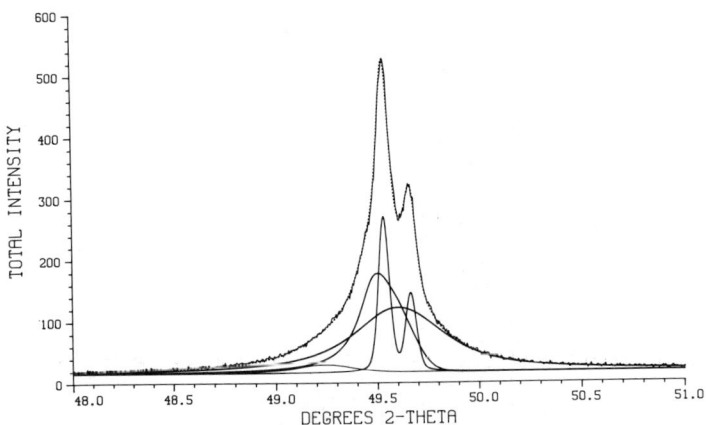

Figure 9. Same hematite 024 peak as in Figure 8 that has been fit with four peaks. The mineral peak shows a clear separation of the Kα1α2 doublet.

distorting the shape of the peak, especially on the upper 2θ side. There is apparently a mineralogical reason why both sides of the experimental profile are distorted. In this case, peaks were added to the experimental profile until a near perfect fit was achieved (Figure 9) in which the peak that is recognized to be the mineral peak has an FWHM of 0.0633 °2θ. The Kα doublet is completely resolved for this peak. In this example, both instrumental aberrations and what are apparently crystal lattice defects are accounted for by the three broad, low-intensity peaks that were used to fit the pattern.

In order to infer the significance of the peaks used for the fit in Figure 9, a different "standard" substance would need to be analyzed with the same diffractometer configuration at approximately the same diffraction angle. The peak for that substance would then be curve-fit and the parameters found for the instrumental peak(s) would be noted. Then, by fixing the parameters for the instrumental peaks found for the "standard" substance, the hematite 024 profile in Figure 9 could be curve fit. All other peaks required to fit around the fixed instrumental peaks could then be attributed to the mineral. The significance of the mineral peaks would then be the subject of another study.

Because of the uncertainty in determining true mineral peak breadths, another approach was used to approximate mineral peak profiles. Figure 10 is a portion of a diffraction pattern from a Hawaiian bauxite. The pattern was collected with diffractometer configuration 1 using CoKα radiation in place of Cu and the divergence slit was fixed at a 1° aperture. The cobalt tube was a long, fine-focusing type operated at 40 kV and 25 mA. The features in Figure 10 are gibbsite 110 and 200 peaks and the goethite 110 peak. For a best fit, only one "instrumental" peak was used which is the broad, low intensity peak below the gibbsite 110 peak. In the first approximation of curve fitting, and with no prior information about the physical state of the two minerals involved, Figure 10 yielded the best peak parameter array that could be extracted from the data. The same specimen was then analyzed by synchrotron powder diffraction, Figure 11.

The physical arrangement of the synchrotron powder diffractometer is such that axial divergence is almost completely eliminated, flat specimen effects and specimen transparency are greatly minimized, the wavelength distribution of the source is much sharper than a standard laboratory X-ray tube, and there is no Kα1α2 doublet (Cox, 1992). The powder pattern in Figure 11 is not completely void of instrumental aberrations but in comparison with the pattern in Figure 10, the effect of the instrument is greatly reduced. A major difference between the powder patterns in Figures 10 and 11 is the wavelength that was used to collect the two patterns. The wavelength of CoKα1 is 1.788965 Å (White and Johnson, 1970). The wavelength of the synchrotron radiation used to collect the pattern in Figure 11 was 1.15014 Å. Therefore, the peak positions in Figure 11 are displaced from those of Figure 10 on the 2θ scale. Also, the mass absorption coefficients for aluminum and iron were different which accounts for the lower intensity of the goethite peak as compared to the gibbsite peaks.

Taking into account the differences in the X-ray wavelength used to collect the patterns in Figures 10 and 11, the peak widths found by fitting the synchrotron powder pattern were converted to widths based on the wavelength of cobalt radiation. Assuming that the converted peak breadths were the true mineral breadths, the FWHM's and positions of the three peaks were fixed (not allowed to vary during curve fitting). Then, "instrumental" peaks that were allowed to vary during curve fitting were added to each of the three mineral peaks The results are shown in Figure 12. The shaded peak areas in Figure 12 are the width-corrected peaks from Figure 11 whereas the three unshaded peaks account for instrumental aberrations. With some doubt still remaining as to the validity of the breadths and shapes of the mineral contributions, we are sure that what Figure

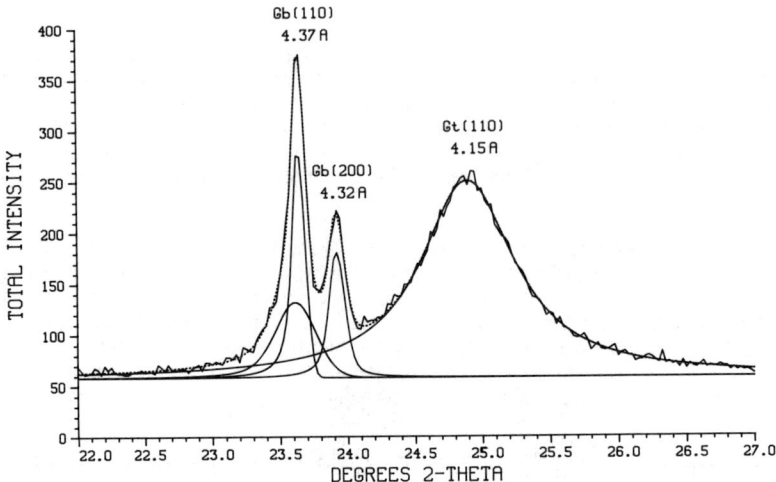

Figure 10. Truncated portion of a diffraction pattern of a Hawaiian bauxite. The pattern was collected by a standard laboratory diffractometer using CoKα radiation and a fixed, 1° divergence slit.

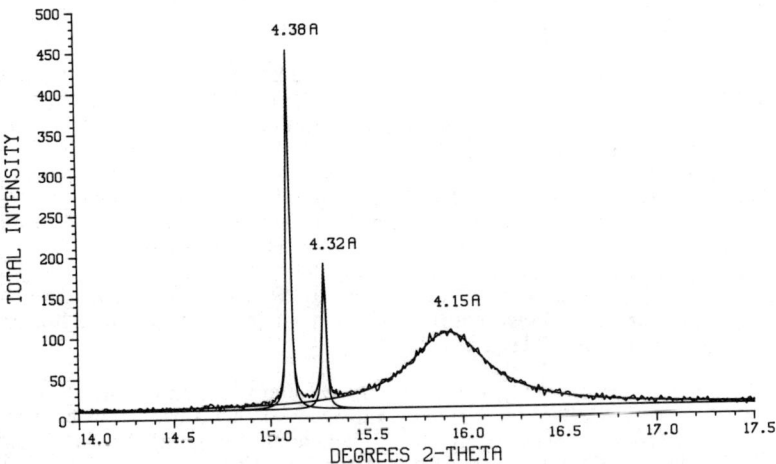

Figure 11. Truncated portion of a diffraction pattern of the same Hawaiian bauxite as shown in Figure 10. This pattern was collected on beam line X7A at the National Synchrotron Light Source, Brookhaven National Laboratory, Long Island, New York.

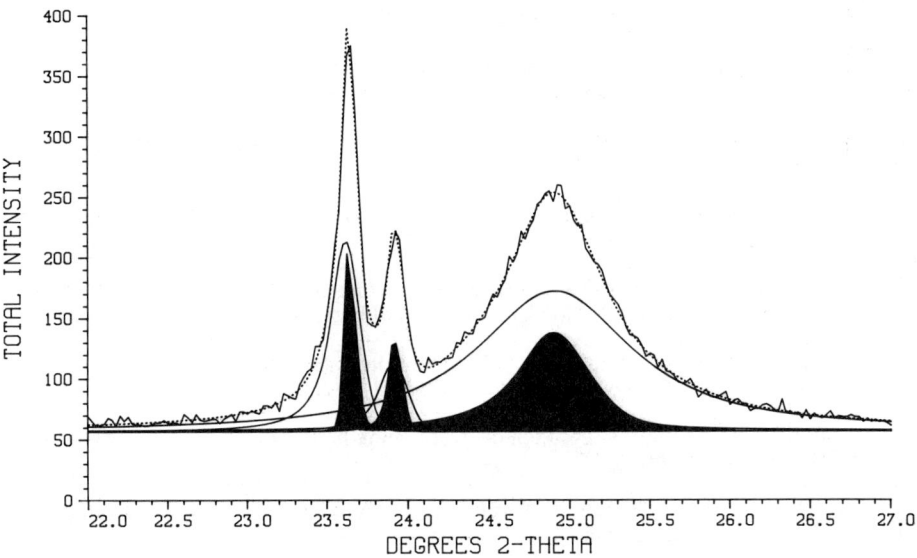

Figure 12. A refitting of the same pattern in Figure 10. The shaded peaks are from the synchrotron powder pattern in Figure 11 that have been corrected for the wavelength and peak position of a diffraction pattern produced by Co radiation.

12 actually shows is the difference between the instrumental aberrations of a laboratory diffractometer and a synchrotron powder diffractometer of the type described by Cox (1992) at the National Synchrotron Light Source, Brookhaven National Laboratory. Synchrotron powder diffraction appears to offer the best prospects for the characterization of true physical mineral diffraction profiles.

CONCLUSIONS

Although unique mineral profiles may not be possible to characterize by curve fitting/peak decomposition, employing a fixed set of diffractometer operating parameters will guarantee that instrumental aberrations will remain constant at a set angular position. Changes in the mineral peak profiles can be quantitatively measured by curve fitting by applying a constant instrumental profile to the experimental profile. We have shown that with synchrotron powder diffraction, experimental peak profiles can be curve fit to as near to the actual mineral profiles as modern powder diffraction methods can achieve. There is no question that there are instrumental aberrations in synchrotron powder diffraction but the magnitude of the aberrations is so much smaller than for standard laboratory diffractometers that the true physical mineral profiles are approached. To assume that a mineral free of diffracting domain size broadening and lattice strains will exhibit diminishing peak breadths, even for diamond, is not entirely valid in our opinion. Therefore, although typical experimental profiles exhibit considerable breadth, "typical" minerals should be expected to exhibit a significant contribution to the overall breadth.

REFERENCES CITED

Alexander, L. (1954) The synthesis of X-ray spectrometer line profiles with application to crystallite size measurements: *Journal of Applied Physics* **25**, 155-161.

Balzar, D. (1992) Profile fitting of X-ray diffraction lines and Fourier analysis of broadening: *Journal of Applied Crystallography* **25**, 559-570.

Bevington, P. R. (1969) *Data Reduction and Error Analysis for the Physical Sciences*: McGraw-Hill, New York, 336 pp.

Cox, D. E. (1992) High-resolution powder diffraction and structure determination: in *Synchrotron Radiation Crystallography*, by Philip Coppens, Academic Press, New York, 186-254 pp.

Ergun, S. (1968) Direct method for unfolding convolution products-Its application to X-ray scattering intensities: *Journal of Applied Crystallography* **1**, 19-23.

Fraser, R. D. P. and Suzuki, E. (1969) Resolution of overlapping bands: Functions for simulating band shapes: *Analytical Chemistry* **41**, 37-39.

Fraser, R. D. P. and Suzuki, E. (1970) Biological applications: in *Spectral Analysis: Methods and Techniques*, J. A. Blackburn, ed., Marcel Dekker, New York, 289 pp.

Henke, B. L., Yamada, H. L. and Tanaka, T. J. (1983) Pulsed plasma source spectrometry in the 80-8000eV X-ray region: *Review of Scientific Instruments* **54**, 1311-1330.

Huang, T. C. and Parrish, W. (1975) Accurate and rapid reduction of experimental X-ray data: *Applied Physics Letters* **27**, 123-124.

Jones, R. C. (1989) A Computer Technique for X-ray Diffraction Curve Fitting/Peak Decomposition: in *CMS Workshop Lectures, Vol. 1, Quantitative Mineral Analysis of Clays*, D. R. Pevear and F. A. Mumpton, eds., The Clay Minerals Society, Evergreen, CO. 52-101.

Klug, H. P. and Alexander, L. E. (1974) *X-ray Diffraction Procedures for Polycrystalline and Amorphous Materials*, 2nd Ed., John Wiley & Sons, New York, 966 pp.

Langford, J. I. (1968) The variance and other measures of line broadening in powder diffractometry. I. Practical considerations: *Journal of Applied Crystallography* **1**, 48-59.

Langford, J. I., Louër, D., Sonneveld, E. J. and Visser, J. W. (1986) Applications of total pattern fitting to the study of crystallite size and strain on zinc oxide powder: *Powder Diffraction* **1**, 211-221.

Marquardt, W. D. (1963) An algorithm for least-squares estimation of nonlinear parameters: *Journal of the Society of Industrial Applied Mathematics* **11**, 431-441.

Post, J. L. and Plummer, C. C. (1972) The chlorite series of Flagstaff Hill area, California: A preliminary investigation: *Clays & Clay Minerals* **20**, 271-283.

Wertheim, G. K., Butler, M. A., West, K. W. and Buchanan, D. N. E. (1974) Determination of the Gaussian and Lorentzian content of experimental line shapes: *Review of Scientific Instruments* **45**, 1369-1371.

White, E. W. and Johnson, G. G., Jr. (1970) *X-ray Emission and Absorption Wavelengths and Two-Theta Tables*, 2nd. Ed., American Society for Testing and Materials, Philadelphia, PA, 293 pp.

Young, R. A. and Wiles, D. B. (1982) Profile shape functions in Rietveld refinements: *Journal of Applied Crystallography* **15**, 430-438.